Antigravity Droplet Surface Materials

David J. Fisher

Published by **Materials Research Forum LLC**
Millersville, PA 17551, USA

Published as part of the book series
Materials Research Foundations
Volume 187 (2026)
ISSN 2471-8890 (Print)
ISSN 2471-8904 (Online)

Print ISBN 978-1-64490-388-9
ePDF ISBN 978-1-64490-389-6

Distributed worldwide by

Materials Research Forum LLC
105 Springdale Lane
Millersville, PA 17551
USA
https://mrforum.com

Printed in the United States of America
10 9 8 7 6 5 4 3 2 1

Table of Contents

Introduction

Superwicking surfaces, known for their ability to transport liquids automatically, have been widely explored in fields such as thermal management, water collection, microfluidics and biomedical devices. This property is inspired by natural examples such as the Sarracenia trichome and spider silk, which use micro/nanostructures to collect and move water efficiently. The performance of such surfaces depends on factors like surface energy, liquid viscosity, and surface morphology. Various fabrication techniques—plasma treatment, chemical or mechanical etching, and especially laser ablation—are used to modify surface wettability. Laser ablation offers advantages such as high flexibility, simplicity, and compatibility with diverse materials. Studies using femtosecond lasers have successfully produced microgrooved or nanopatterned structures that enhance wicking performance, although the relationship between surface design and wicking dynamics remains under-explored.

Inspired by natural materials such as cactus spines, researchers have developed artificial materials capable of directional liquid transport. Such motion is primarily driven by gradients in wettability or surface topography, though challenges remain due to adhesive resistance. Recent advances have focused on optimizing microstructures to improve self-propelled and long-distance fluid transport. Passive mechanisms based upon Laplace-pressure gradients, as seen in spider silk and pitcher plants, have enabled controlled droplet motion without requiring external energy. Despite progress, issues like imprecise droplet trajectories and fluid loss during passive transport persist. Understanding and mimicking these bioinspired strategies offer pathways to developing efficient liquid-handling systems for applications in microfluidics, oil transport and energy-efficient surface design.

Recent progress in microfluidics, enabling precise detection using minimal sample volumes, has driven the creation of advanced chemical and biological fluid analysis platforms such as lab-on-a-chip and miniaturized total analysis systems. These technologies have broad applications in high-throughput screening, diagnostic assays, cell biology and drug delivery. Compared to traditional benchtop laboratory equipment, microfluidic systems offer the advantages of automation and rapid processing. Fluid movement in microscale channels moreover increases the likelihood of bioconjugation between target substances and sensor-immobilized bio-recognition agents.

Microfluidic technologies were initially fabricated using materials suitable for controlled laboratory conditions to study electrophoretic phenomena, cell sorting and micro-patch clamping. Early systems employed rigid substrates such as silicon and glass, which were patterned and etched via clean-room lithography, and then interfaced with pumps and valve components. The brittleness of these materials, together with their complex fabrication and sealing requirements however limited large-scale application. The introduction of polydimethylsiloxane was a turning-point in the field. It became widely used due to its optical clarity, elasticity and strong bonding to materials such as glass and plastic. In spite of these advantages, polydimethylsiloxane-based microfluidics suffered from problems of scalability to mass production. It could also leach unreacted oligomers and absorb small molecules, thus compromising chemical and biological sensing reliability. The reliance on external pumps, tubing and non-degradable polydimethylsiloxane components also restricted its suitability for disposable or in-field sensing.

Rapid advances in micro/nano-technology and surface engineering have accelerated the development of responsive and adaptive surfaces. Droplet and bubble manipulation on such surfaces can be achieved by constructing gradients in wettability, adhesion, or temperature in response to stimuli such as heat, air-pressure, magnetic fields, electric fields or light. Among these stimuli, light-induced photothermal effects offer particular benefits, such as precise spatial control, non-contact actuation and instantaneous response.

Design of an effective slippery photothermal surface for droplet and bubble manipulation requires two key features: high surface porosity and rapid temperature responsiveness under light irradiation. Some studies typically fabricated these surfaces by infusing lubricants into porous solid substrates by using templates or laser-engraving. These methods produced relatively large pores, thus reducing oil-absorption efficiency. In addition, some proposed photothermal materials, such as graphene, Fe_3O_4, CuS, and heterojunctions exhibited low heating-rates under illumination, resulting in a sluggish temperature-response. Prior research had also demonstrated droplet or bubble movement only on tilted surfaces by modulating surface wettability via light. Simultaneous control of both droplets and bubbles remained difficult. This limitation arose mainly from the instability of lubricated surfaces underwater and the slow photothermal response when submerged. Slippery photothermal-responsive laser-induced graphene/polyimide membranes appear to be promising in this respect. By using a combined electrospinning and laser-ablation approach, a laser-induced graphene/polyimide layer can be formed on a polyimide substrate. This offers an enhanced porosity and a rapid light-induced thermal response for controllable droplet and bubble transport.

When a water droplet is placed on a solid substrate, its edges spread in accord with the Tanner law for small droplets or the Lopez law for larger ones, while the droplet's centre-of-mass remains stationary in the horizontal direction. Over the past decades, numerous studies have explored the motion of entire droplets using both experimental and theoretical methods. It was demonstrated that droplets could move not only across horizontal surfaces but also along inclined substrates. Such spontaneous droplet motion promises application in inkjet-printing, micro/nanofluidics and pulsating heat pipes.

Such movement is driven by a surface-energy gradient that must overcome both drag and contact-angle hysteresis forces. The surface energy depends upon the properties of the substrate as well as upon factors such as temperature. Theory and experiment has shown that droplets can move under a thermal gradient, though the velocity is typically tens to thousands of times lower than when motion is induced by a surface-treatment gradient. Various phenomena can create gradients, with chemical processes being the most common although geometrical patterning and photodegradation also produce useful results. In particular, an ultraviolet-sensitive coating which is exposed along a specific direction can generate a gradient. Modification of surface roughness has also been essayed, although this often increases contact-angle hysteresis.

Some have theoretically estimated the maximum velocity which is attainable by a droplet that is elongated perpendicular to its motion, while others have developed a more comprehensive theory which was based upon the total footprint-area in order to describe droplet-velocity changes. On horizontal treated substrates, droplet velocities are typically below 10mm/s. Others demonstrated that droplets could even climb uphill on surfaces which were inclined at 30°, although the maximum velocity decreased from 42mm/s on a horizontal wafer to 18mm/s at the above angle due to gravitational force.

Controlling surface-wettability is critical to enabling precise unconventional droplet behaviour in scientific and industrial fields. Surfaces having extreme wettability, superhydrophobic or superhydrophilic, attract great interest with regard to engineering application. That is, superhydrophobic surfaces with contact-angles greater than 150° are used for self-cleaning and anti-corrosion coatings, while superhydrophilic surfaces with contact-angles of less than 10° are used in anti-fogging, adhesion-enhancement and the suppression of droplet formation. Traditional wettability-control techniques suffer however from limitations in applications which require droplet-jumping, pattern-formation or dynamic motion.

Superhydrophobicity is generally defined as being that property of a surface in which water droplets take up an apparent contact-angle greater than 150°, together with a contact-angle hysteresis of less than 10° and a sliding-angle of less than 5°. It is a property which occurs in Nature, and a great deal of effort has been aimed at mimetically exploiting the superhydrophobicity exhibited, for example, by the lotus leaf.

The fact that the droplets are very far from wetting the surface upon which they stand leads on to many associated tendencies, such as impeding fogging, icing and corrosion. The aim of the present work is to cover the ways in which superhydrophobicity has been imparted to metals. Metals themselves tend more naturally to be hydrophilic and so imparting superhydrophobicity relies upon adding some sort of coating.

William Barthlott (1946-), the discoverer of the lotus effect, noticed that two factors were involved in producing the effect. One was the presence of a waxy material, and the other was the existence of numerous microscopic bumps. The hydrophobic nature of the waxy material causes the water-drops to minimize their contact area with the surface by increasing the angle between the water surface and the leaf surface. The bumps then push the behaviour to superhydrophobic levels, as air which is trapped between the bumps increases the contact-angle. Because of the greater contact-angle, the drop becomes almost spherical in shape and rolls off the leaf.

Based upon the contact-angle and the associated wetting behavior, solid surfaces can be grouped into 4 classes: superhydrophilic (contact-angle less than $10°$), hydrophilic (contact-angle between $10°$ and $90°$), hydrophobic (contact-angle between $90°$ and $150°$) and superhydrophobic (contact-angle greater than $150°$). The contact-angle and its hysteresis are the most important parameters involved in judging the degree of superhydrophobicity of a solid surface. The contact-angle is used as a common measure of wettability. Contact-angle hysteresis is a measure of the stickiness of the water drop to the solid surface, and is the difference between the advancing angle and the receding angle. If drops are to roll off very easily, a high contact-angle and a low hysteresis are essential. The contact-angle of a typical hydrophobic solid surface is between $100°$ and $120°$ and can be increased by increasing the surface roughness.

Superhydrophobicity is often explained by invoking the apparently unrelated Leidenfrost effect, in which water drops on a hot solid surface run around freely. This is because a film of vapour is formed between the surface and the drop, and the latter never makes contact with the solid surface. The Leidenfrost effect is possible if the apparent contact-angle is $180°$, hence the analogy with superhydrophobicity. The capillary length of water is about 2.7mm under ambient conditions. If the drop is smaller than the capillary length, it is considered to be small and, in that case, the gravitational force can be neglected and surface tension predominates. When a small drop is placed on an ideal horizontal solid surface, three interfaces exist: solid/liquid, liquid/vapour and vapour/solid.

Overall, the creation of superhydrophobic surfaces requires a combination of surface roughness creation and surface free energy reduction. The wettability of a solid surface is determined both by the microscopic geometry and by the chemical composition of the surface. Liquid droplets contact the substrate so as to form a solid-liquid-gas contact line. When the droplet reaches a steady state on the solid surface, it takes up a certain angle with respect to the surface. At the intersection of the solid-liquid-gas 3-phase contact, the angle between the tangent plane of the liquid-gas boundary and the solid-liquid boundary is defined as the water contact-angle of the droplet on the surface. The value of this angle is an important means for assessing the wettability of solid surfaces. Each has its own associated interfacial surface tension. Thomas Young, of course, established the basic shape of a liquid drop on a solid. The contact-angle for a completely hydrophobic spherical drop is 180° and the angle for a completely hydrophilic wetted surface is 0°. If the surface tensions can be assumed to be material constants, the contact-angle is also a material constant. Young's equation was however derived for smooth surfaces, and not for rough surfaces. The relationship between contact-angle and surface tension was originally studied by Yang, Wenzel and Cassie. Wenzel assumed that the liquid completely wetted a rough surface whereas Cassie and Baxter assumed that a drop simply sat on a rough surface, with air trapped beneath it. The Wenzel model is usually applied to hydrophilic surfaces while the Cassie & Baxter model is more often used to describe the contact-angle of hydrophobic or superhydrophobic surfaces. In the more developed Pillar model the contact-angle depends upon the fractal dimension and the upper and lower limit-lengths of the fractals, while using Wenzel/Cassie states as the basis of analyses. Surfaces with high contact-angle and low sliding-angle have better self-cleaning effects. There are differing views concerning contact-angle hysteresis. Some relate friction at the triple-phase interface to hysteresis of the contact-angle.

The Young equation was modified by Wenzel so as to describe the relationship between roughness and contact-angle. This was done by introducing a so-called roughness factor: the ratio of the actual surface area to the geometrical surface area. The roughness factor is equal to unity for smooth surfaces, and Young's equation is then retrieved. The term, contact-angle, is correct only for smooth surfaces. The term, apparent contact-angle, is applicable to rough surfaces. The Wenzel theory argues that the solid/liquid and solid/vapour interfacial surface tensions are somewhat increased by the increased rough surface area while the liquid/air interfacial surface tension is unchanged. A greater contact-angle is therefore taken up in order to balance the increased surface tension. Even if the smooth solid surface is hydrophilic, the roughened solid surface can become superhydrophilic. If the smooth surface is hydrophobic, the roughened surface can become superhydrophobic. The Wenzel theory was further extended by Cassie and Baxter for the case of porous heterogeneous surfaces. When the surface roughness is higher, it is not essential that the liquid should fill the entire solid surface. The liquid need contact only the peaks of the surface and not enter the valleys. While the bulk of the liquid contacts the peaks of the surface constituting the solid/liquid interface, the remainder contacts any vapour present in the valleys of the solid surface constituting the liquid/vapour interface. Superhydrophobicity is imparted by a suitable choice of roughness, surface texture and added low surface-energy materials.

As noted, there are two main forms of wetting of a rough solid surface: Wenzel and Cassie-Baxter, the modifications of the Young model for smooth surfaces. The Wenzel model assumes that water droplets will fully infiltrate rough surface structures. The Cassie-Baxter model assumes that the contact-angle will increase when there are microstructures on the surface because air is trapped in any gaps so as to form an air cushion which supports the water droplets. When the solid surface is hydrophobic, the static contact-angle alone is not adequate for fully describing the wettability. It is essential to measure the dynamic contact-angle. As the volume of a droplet is increased, the contact-angle also increases and the contact boundary of the solid/liquid interface tends to advance. Above a threshold size, the contact boundary of the solid-liquid interface moves outwards, with the corresponding contact-angle being termed the advancing contact-angle. As the volume of a droplet is decreased, the contact boundary of the solid/liquid interface tends to recede. When the volume of the droplet decreases to a certain threshold, the contact boundary of the solid/liquid interface of the droplet moves inwards, with the corresponding contact-angle being termed the receding contact-angle. The contact-angle hysteresis is then the difference between the advancing contact-angle and the receding contact-angle. The smaller the contact-angle hysteresis, the easier it is for the liquid to leave the surface.

The various methods used to obtain superhydrophobic surfaces can be divided into two main types: top-down and bottom-up. In the former case, the required surface is obtained by etching. In the latter case, the same effect is obtained by chemical deposition. A combination of the two methods can also be used. If the treated surface is not then superhydrophobic, post-treatment with hydrophobic materials can be used to obtain superhydrophobicity. Common hydrophobic materials include silanes, and fluorinated or hydrocarbon thiols. The inclusion of micro-fibres or nano-fibres can provide a good surface texture for water-repellent surfaces.

If a surface can be created with a very low fraction of air at the nano-scale, superhydrophobic surfaces can be produced even from hydrophilic materials. The bottom-up approach is good for preparing very low air fraction superhydrophobic surfaces. The bottom-up methods include chemical vapour deposition, electrochemical deposition and layer-by-layer deposition. Electrochemical processes are capable of controlling both the surface roughness and the surface morphology. Common electrochemical processes are anodization, electrodeposition of conductive polymers, electrodeposition of metals and metal oxides, and electroless galvanic deposition. The anodization creates a porous nanostructured oxide layer. In electroless galvanic deposition, spontaneous deposition of metallic ions occurs when they contact a metallic surface with a lower oxidation potential. The advantages of the bottom-up and top-down approaches can be combined to produce superhydrophobic surfaces with a two-scale roughness. A micro-scale rough surface is first produced by using the top-down approach, and nano-scale roughness is then added by using the bottom-up approach.

Anti-icing is a case in which superhydrophobicity can be effective. Ice adhesion can impair the aerodynamics of aircraft. The superhydrophobic properties of surfaces weaken the adhesion between ice and surface, leading to easy removal by normal or shear forces. A higher contact-angle and a low contact-angle hysteresis are responsible for normal and shear forces, respectively. Icephobic surfaces have a shear strength of between 150 and 500kPa. Superhydrophobic surfaces are sometimes ineffective as anti-icing materials. Ice-accretion on the surfaces can be delayed, when compared with that of flat hydrophobic surfaces, but gradually damages the surface microstructure during icing and de-icing, thus reducing the anti-icing properties. Adhesion-strength increases however, in a humid environment at low temperatures, due to condensation and an associated anchoring effect. The superhydrophobicity prevents icing by forcing water to merge into large drops which then roll off a surface before freezing can occur. The formation of ice can be quite complicated and may involve the formation of quite exotic shapes[1].

True self-cleaning surfaces are those which combine superhydrophilicity and photocatalysis to break down dirt and wash it away. Superhydrophobic surfaces are extremely dry, and repel water drops. So these surfaces do not in fact clean themselves but, when water drops roll over the surface, they wash away dirt. Raindrops should fall at high speed in order that dust particles be washed away. Corrosion resistance, and drag-reduction in underwater applications, are other beneficial effects of superhydrophobicity. When a superhydrophobic surface is completely immersed in water the entrapped air is separated from the moving water. If the air pockets cover a sufficiently large area, the superhydrophobicity can reduce skin friction and cause a slip effect. The degree of superhydrophobicity is decreased if the amount of entrapped air is reduced. It is therefore critical to maintain air pockets in underwater applications.

Biological fouling, also known as biofouling, is the accumulation of the biological matter on a solid surface, together with deposits of corrosion, ice and suspended particles. If a superhydrophobic surface present, it reduces the contact area between water and a solid surface, thus restricting the amount of biological matter reaching the surface. Avoidance

Materials Research Forum LLC
https://doi.org/10.21741/9781644903896

of biofouling is possible only if air pockets can be stabilized within pores so as to prevent biological matter from adhering to the solid surface. Since the volume of water is so large, biofouling cannot be characterized by an apparent contact-angle, contact-angle hysteresis or sliding angle. The fraction of wetted area is the only available measure of biofouling superhydrophobicity. Self-propulsion of droplets can occur on a superhydrophobic plant surface such as a lotus leaf. Self-propulsion releases the droplet from the surface/air interface, and any associated adhesion, and leaves it exposed to any forces that can perhaps transport the droplet over significant distances. Gravity and air currents can lead to total removal of the propelled droplets. This can also remove contaminant particles from the surface if they are similar in size to the water droplets[2].

A quite spectacular phenomenon can occur upon letting a water droplet fall onto a pore in a superhydrophobic plate on a water surface. There can be a spontaneous transformation of surface energy into gravitational potential energy[3]. On the basis of this self-capturing phenomenon, a power-free water-pump was created[4] which comprised a superhydrophobic plate with a pore, mounted on a leak-proof cylindrical container filled with water. This led to the anti-gravity long-distance transport of water. The use of a superhydrophobic surface, having the ability to withstand high pressures and exhibiting low adhesion, could constitute a power-free pump. The lifting-height of the pump was of the order of 100mm, and increased with decreasing pore diameter. The transport capacity of the pump was unaffected by the tilt of the pump body or by the tube diameter. Continuous delivery of water was possible over distances of the order of metres.

The oxide films on aluminium surfaces easily absorb oil and other pollutants, causing damage to the surfaces and reducing the service life of the material. A superhydrophobic surface offers self-cleaning, anti-icing, anti-corrosion and anti-fouling properties, but suffers from poor mechanical durability. It is therefore important to improve that durability. Increasing the substrate roughness is one of the available means for improving the durability of a superhydrophobic surface, in that an increase in the substrate roughness changes the microstructure and thus influences the mechanical behaviour. The roughness can be increased by laser-processing, chemical etching and sanding. Sand-blasting is widely used to clean surfaces and create rough surfaces, but its use to create a rough substrate for superhydrophobic surfaces is still being explored.

The methods which are used to produce a superhydrophobic metal surface can also be divided into direct and indirect. The latter involves the creation of a superhydrophobic coating on the metal surface. Coatings can greatly improve the superhydrophobicity, corrosion resistance and other properties of metal surfaces but they cannot change the wettability of the metal itself. Poor adhesion between a coating and a metal substrate also limits their service life. The direct method instead involves a process which endows the metal surface itself with superhydrophobicity, although this may lead to considerable loss of material.

For the precise and active manipulation of droplets in fields such as microfluidics, biomedical chemistry, cell-culture and energy systems, surfaces which offer a spatially-varying polarity are used to guide droplet motion. The obtention of such surfaces usually calls for hierarchical micro/nanostructures, but the integration of structures having different extreme wettabilities on the same surface remains challenging due to differences in fabrication methods and specification.

The wetting behaviour of solids, usually characterized via contact-angle measurements, provides an insight into adhesion, surface topography and composition. Theoretical and practical aspects of contact-angle analysis are well-established, especially high-precision drop-shape analysis. Determination of the advancing and receding contact-angles, which differ due to contact-angle hysteresis, usually involves visual observation of the triple-line motion. This process is however highly dependent upon experimental conditions, observational subjectivity and other factors which lead to inconsistent and irreproducible results: even measurements which are performed on identical surfaces may vary significantly. High-precision drop-shape analysis improves reproducibility by tracking triple-line motion, permitting detailed characterization of contact-angle behaviour. This permits data-fitting using sigmoidal functions which effectively describe the overall contact-angle behavior, especially on smooth surfaces, and help to visualize surface effects and define specific angles.

In one seminal study, a surface having a chemical gradient produced by chemical vapour deposition permitted 1 to 2µL water droplets to climb an incline of 15°. Other work further extended this principle by dissolving surfactant additives in decane, and this allowed droplets to climb inclines of up to 43°. Although effective for oil droplets, chemisorption-based methods are less suitable for water droplets. This is due to the limited concentration of free surfactants in water; particularly those containing highly hydrophobic perfluoroalkyl chains.

In order to overcome this limitation, a self-propelling system which used the hydrofluoric acid etching of silicon surfaces was introduced. This reduced the surface free-energy without encountering the solubility problems which were associated with surfactants. This then enabled the spontaneous motion of water droplets. The system allows for significant uphill motion and provides insights into the relationship between reaction-rate, droplet-velocity and ambient temperature and offered a new mode of chemo–mechanical energy conversion.

Materials Research Forum LLC
https://doi.org/10.21741/9781644903896

Directional propulsion of droplets on heated surfaces is of great practical value for applications in power generation, drag-reduction, microfluidic devices and spray-cooling. In high-temperature environments, controlled droplet motion towards hot-spots increases heat-transfer and cooling efficiency. Directional motion requires the breaking of surface symmetry in order to generate a driving force which is sufficient to overcome contact-line pinning. This can be achieved by providing asymmetrical surface geometries, wettability-gradients or external stimuli such as light, electric or magnetic fields.

When the surface temperature exceeds the Leidenfrost-point, a vapour-layer forms beneath the droplet; the Leidenfrost effect. This layer minimizes contact-line friction and allows droplets to move under very little driving force; often less than 10µN. In order to exploit this effect, asymmetrical structures such as vertical or planar ratchets, gradient micro-pillars and non-parallel plates have been developed. These serve to direct droplet-motion via asymmetrical vapour flow. The vapour-layer also introduces however a thermal resistance which limits the heat-transfer efficiency. That is, in spray-cooling, droplet impacts produce rapid heat-transfer via violent evaporation. but the uncontrolled motion can reduce overall efficiency. Ensuring effective and directional droplet-propulsion near to the Leidenfrost-point is essential for maximizing the heat-transfer performance at high temperatures.

Precise manipulation of very small liquid volumes on solid surfaces is thus critically important. Nature presents many examples of surfaces which can actively control droplet motion. Key problems remain unresolved, including a lack of durability, difficulty in reconfiguring surfaces, fabrication difficulties, slow response times, significant contact-angle hysteresis, a limited ability to manipulate droplets across large areas and the risk of contaminating the liquid droplet or substrate. It has been noted that many existing strategies rely upon wettability gradients generated by physical patterning, chemical patterning or both. Most approaches involve a permanent surface modification which is tailored in order to accommodate a single function, thus limiting their usefulness in reconfigurable systems. A more versatile strategy is to incorporate a stimulus-responsive behaviour into either the liquid or the substrate. Modifying the liquid can cause unwanted contamination, and a large contact-angle hysteresis restricts applicability to certain liquids, while response-times are often slow. Optical control of oil droplets, but not water droplets, has been achieved for speeds of a few micrometres per second. Water droplet motion over a TiO_2-nanorod coated polymer surface at 1mm/s was possible but required 90 minutes of ultra-violet pre-irradiation, plus and several days in the dark in order to reverse fully the wettability. Again, in surfaces in the natural world play a critical role in mass and energy exchange in biological and non-living systems. As noted above two types of interfacial material, slippery lubricant-infused porous surfaces and superhydrophobic surfaces, have attracted attention for use in self-cleaning, drop-condensation, anti-icing and anti-fouling purposes. Replacing air-pockets with a lubricant layer creates those properties; even under harsh conditions. Lubricant layers also have drawbacks however. They mask the underlying solid surface and impair effects arising from structural gradients or surface-charge differences. They also hinder stimulus-responsive droplet manipulation. Current systems depend upon tunable substrates or changeable lubricant layers, and typically require gravity in order to overcome contact-line pinning. Droplet-transport tends to be slow, short-range and difficult to control. External stimuli such as light, magnetic forces or electric fields could improve that motion, but it often remains limited and contamination-prone.

In order to offset these limitations, another type of slippery material was developed which combined the benefits of solids and lubricants, and provided light-induced surface-charge regeneration and well-controlled droplet motion under diverse conditions. Unlike the original surfaces, the light-induced charged slippery surfaces incorporate three main components: gallium-indium liquid-metal micro-particles for converting light into localized heat, co-polymer which offered marked ferroelectric behaviour and a micro-structured layer which was coated with hydrophobic silica nanoparticles that secreted the lubricant. As compared with previous photo-controlled systems, this material enables significantly faster (18.5mm/s) droplet transport over much longer (100mm) distances against gravity for droplet volumes ranging from 0,001 to 1500μL.

Bio-inspired designs, based upon spider-silk or cactus-spines, have produced artificial materials capable of guiding droplets. Examples include conical fibres, asymmetrical slippery surfaces, chemical gradients and smart adhesive materials. Nearly all of these studies focus on open surfaces. Transport within tubes is also important because it minimizes evaporation and contamination. Most research on tubular systems prioritizes anti-fouling behaviour via superwetting coatings rather than actively controlling droplet-speed or direction. Some research has led to limited control. For example, improved water self-siphoning has been obtained by mimicking peristome structures on tube interiors, although fluid-loss is unavoidable. Directional transport has been achieved by inducing asymmetrical deformation in a light-responsive liquid-crystal polymer tube. Similar strategies have been applied to magnetic polydimethylsiloxane tubes. Responsive molecules have been added to the inner surface in order to create contact-angle gradients and procure higher-speed water-transport. All of these approaches require external mechanical input, and the shape reverts when the force is removed. External force can be undesirable. Due to the intrinsic hydrophobic nature of polydimethylsiloxane, only low surface-tension liquids can be transported, unless additional coatings are applied.

Another strategy is to combine a shape-memory polymer tube with a pitcher-plant inspired slippery inner surface. The shape-memory effect permits the tube geometry to be tuned, while the slippery coating reduces droplet adhesion and widens the range of usable liquids: both water and oils having various surface tensions can be controllably transported. The speed and direction can be adjusted, and multiple transport modes are possible. A 15µL water droplet which is introduced at one end can move to the other end of its own accord. If the tube-shape is changed from round to tabular, heating below the droplet can reverse the direction of motion and enable return travel. This flexibility exceeds that of systems having fixed gradients of chemistry or geometry. Asymmetrical deformation and a slippery coating are essential as, without them, self-transport does not occur.

The rapid removal of condensed droplets from surfaces not only permits self-cleaning, increased condensation and anti-icing, but also leads to heat-transfer coefficients which are up to 30% higher than traditional drop-wise condensation surfaces. For efficient condensate removal, the surfaces must exhibit superhydrophobicity; leading to unstable three-phase contact lines on textured surfaces. Self-cleaning and condensate-repelling surfaces have been fabricated by using nano- or micro/nanostructures. Although nanoscale features are considered essential, they can trap tiny droplets and provoke a transition from the Cassie to the Wenzel state, thereby increasing surface adhesion and potentially leading to flooding. The choice of a surface morphology which permits self-removal is a problem because trapped condensates have to move uphill within surface-textures in order to prevent a wettability transition and maintain superhydrophobicity.

Magnetically responsive superhydrophobic surfaces which are capable of tunable wettability have been produced which offer the precise control of droplet behaviour. Such magnetic actuation is especially appealing due to its characteristics of remote, rapid and non-invasive control. Magnetic surfaces can guide droplets for purposes such as cell-culture, energy-harvesting and controlled chemical reaction.

Tunable super-wetting surfaces that can alternate between Cassie-Baxter (low-adhesion) and Wenzel (high-adhesion) states are of great interest, but maintaining reversible wettability control is difficult. Developments in lubricant-infused porous surfaces, using materials such as ferrofluids, paraffin and ionic liquids, are promising for achieving stable switchable wettability.

Self-propelled droplet motion, driven by surface-energy gradients, is an efficient method for converting surface energy into mechanical energy. Droplet movement due to chemical, thermal or photo-induced gradients often requires external energy sources. A simpler approach involves creating a shape and wettability gradient by using hydrophilic and hydrophobic materials, thus permitting droplets to move autonomously. Factors such as the gradient angle, droplet volume and contact-angle hysteresis control motion-efficiency. Under certain vibration conditions, droplets can climb inclined surfaces, due to non-linear friction and contact-angle dynamics.

A liquid droplet which is resting on a solid surface experiences retention-forces which arise from hysteresis in its static contact-angles. When the surface is tilted, the droplet deforms but remains stationary until the contact-angles at the front and rear exceed the advancing and receding contact-angles, respectively. This hysteresis originates from surface imperfections, which make the displacement of the contact-line energetically unfavourable. When the hysteresis is sufficiently small, the droplet begins to slide at moderate inclination angles and often transform into strongly asymmetrical forms. When the hysteresis is large, the droplet can remain pinned to the substrate even when it is vertical. This pinning is a problem for microfluidic applications and motivates a search for mechanisms which can overcome hysteresis. These include the use of wettability-gradients, the coupling of thermal effects with ratcheting and the use of asymmetrical vibrations.

By analogy with the manner in which an object can overcome friction by substrate oscillation, it is reasonable to expect that a sessile droplet on an incline will slide downward if the incline is subjected to vertical vibrations. Results demonstrate that a droplet which is placed on a vertically vibrated inclined plane can actually ascend the surface, provided that the acceleration amplitude is sufficiently high. If a substrate is mounted on a shaker which imposes sinusoidal vibrations, the key control-parameter is the maximum acceleration. Under partial wetting conditions, a droplet having a characteristic size which is comparable to the capillary length exhibits a rocking motion about its equilibrium position. The vibration frequencies which are used correspond to the droplet's first resonant modes, ranging from 30 to 200Hz for droplets of 2 to 10μL.

A droplet resting on a stationary inclined plane either remains fixed or slides downward. A stationary state occurs only when the contact-angle hysteresis is sufficiently strong. It might be expected that when the plane is vibrated, a sliding droplet continues to slide while a pinned one stays in place if the vibration is weak or begins to move when the vibration amplitude is large enough. This intuition is mistaken in that droplets on a vertically oscillating inclined plate can migrate upward against gravity. When droplets are placed on a horizontal substrate which is subjected to both vertical and horizontal oscillations and the amplitude-ratio and phase-difference between the two oscillations is varied, the mean droplet velocity and the direction of motion can be tuned.

Theoretical studies have analysed the dynamics of droplets on vibrating surfaces, including a droplet on a horizontally oscillating plate, by treating it as a forced linear oscillator whose frequency is matched to the droplet's natural oscillation. The results indicate that the direction of droplet drift depends on the anharmonic component of the vibration. Studies have been made of droplets on inclined plates which are vibrated vertically using a quasi-static approximation, assuming that oscillations are sufficiently slow that the droplet shape results from a balance between surface tension, gravity and inertia. The results predict that a droplet will climb uphill if the substrate acceleration profile contains sharp narrow troughs and broad shallow crests. Study of droplets on a horizontally vibrating plate under tangential harmonic motion shows that the droplets spread symmetrically without undergoing a nett displacement.

The study of vibrating droplets has its roots in classical fluid dynamics and dates back to the works of Kelvin, Rayleigh and Lamb which characterized the vibration-modes of free liquid drops. Later work extended these analyses so as to treat supported nearly-spherical droplets in the absence of gravity, for both cases with pinned contact-lines and those with mobile contact-lines which exhibit hysteresis. A common experimental observation is that sufficiently strong vibrations excite capillary waves on the droplet surface and which, if amplified, can lead to break-up and spray-formation.

The influence of substrate-heterogeneity upon droplet motion under vibration has attracted attention. Carefully designed heterogeneous surfaces can be exploited so as to guide droplet motion using vibrations. Seminal work showed that surface vibrations can help droplets to overcome the energy-barriers which are constituted by surface roughness and chemical heterogeneity. Experiments involving acoustic vibrations of horizontal surfaces confirm that contact-angles vary with vibration-intensity.

In what follows are described the various substrates which have been used to support the anti-gravity motion of droplets, together with the effect of substrate-oscillation ... regardless of the nature of the substrate.

Aluminium

The creation of low surface-energy non-fluorinated dual-scale so-called *Allium giganteum* structured superhydrophobic surfaces on 5N-purity aluminium was carried out by using a one-step electrodeposition method[5]. The maximum contact-angle of 168.6° was found for a deposition voltage of 30V. Ultra-fast (less than 120s) one-step electrodeposition was used[6] to create a fluorine-free superhydrophobic surface having a dual-scale hierarchical papillae structure having a low surface energy. The as-prepared surfaces exhibited extremely low surface-adhesion, and a self-cleaning capability.

Mirror-finish superhydrophobic 4N-purity aluminium surfaces were prepared[7] via the formation of anodic alumina nanofibres and subsequent modification with self-assembled monolayers. The high-density nanofibres were formed on the surface by anodizing in pyrophosphoric acid solution. The nanofibres became entangled and bundled by further anodization at low temperatures, and the aluminium surface was completely covered in long floppy nanofibres. The surface had a contact-angle of less than 10°. When the nanofibre-covered surface was modified with n-alkylphosphonic acid self-assembled monolayers, the water contact-angle suddenly shifted to superhydrophobic, with a contact-angle greater than 150°. The angle increased with applied voltage during pyrophosphoric acid anodizing, anodizing time and number of carbon atoms in the monolayer molecules on the alumina nanofibres. By optimizing the anodizing and modification conditions, superhydrophobic behaviour could be obtained by using a pyrophosphoric acid anodizing period of just 180s with subsequent immersion.

An alternative method[8] for enhancing the superhydrophobicity of 3N-purity aluminium surfaces led to a contact-angle of about 153°, due to the formation of an hierarchical structure via the grinding and polishing of micro-grooves into the surface, combined with simultaneous exposure to hydrochloric and dodecanoic acids. The metal and its oxide appeared to be involved, with free-aluminium anchored to fatty-acid molecules and to alumina molecules. Metallic aluminium and its oxides were both presumably required to form the compound that was responsible for the superhydrophobicity. An important factor was the existence of features which held air and prevented water droplets from coming into contact with the solid surface. The homogeneous Wenzel model predicted that a water droplet can penetrate asperities, while the composite Cassie–Baxter model suggested that a droplet could be suspended above asperities when gas is trapped in the cavities of a rough surface.

A superhydrophobic coating was created[9] on a 3N-purity aluminium surface by anodization in a sulphuric acid electrolyte, followed by surface modification with myristic acid. The contact-angle of the coatings increased from 114.1° to 155.2° upon increasing the anodization voltage from 0 to 22V. The contact-angle markedly improved when the anodization voltage reached 20V. When the voltage was further increased to 22V, the contact-angle and sliding-angle worsened from 155.2° and 3.5 to 152.8 and 7.0 , respectively. The as-prepared coating had an hierarchical micro-nano structure, with a static water contact-angle of 155.2° and a sliding-angle of 3.5 . It retained a contact-angle of up to 151.1° following sand-blasting for 60s and remained stable after exposure to acidic and alkaline solutions. Superhydrophobic 3N-purity aluminium with an hierarchical micro-nano structure was produced[10] by combining anodization, chemical etching and surface-modification involving polydimethylsiloxane coating via chemical vapour deposition. The anodizing produced a cylindrical nanopore (70 to 90nm) network. During etching, the diameters of the pores increased and the surface modification deposited SiO_2-polydimethylsiloxane nanoparticles over them. Anodized aluminium, following modification, became superhydrophobic with contact-angles of 153° and 154°, respectively, with a sliding-angle of less than 1°. The charge-transfer resistance of the superhydrophobic surface of the modified anodized-etched material was $16700k\Omega cm^2$. This was some 900 times higher than that of pure aluminium.

The hierarchical growth of a γ-AlOOH film on 2N5-purity aluminium foil was achieved[11] by using solution-phase synthesis. The resultant film surface consisted of 3-dimensional micro-protrusions which were made up of well-aligned nano-needles and were a biomimetic version of lotus leaves. The surface, following hydrophobization, had a water contact-angle of 169° and a sliding-angle of about 4° for 5µL droplets. This was attributed to the combined effects of a dual-scale roughness at the micro- and nanometre levels. The film exhibited a relatively good adhesion to the aluminium substrate, and retained superhydrophobicity following ultrasonic treatment. There was a partial loss of superhydrophobicity following abrasion.

Chemical etching produced[12] superhydrophobic aluminium surfaces having a water-contact angle of 154.8° and a sliding-angle of about 5°. The etched surfaces exhibited irregular micro-scale plateaux and hollows within which there were nano-scale block-like convex and hollow features. Superhydrophobicity existed only for structures in which the plateaux and caves were suitably ordered.

Superhydrophobic surfaces were created[13] on 100µm pure aluminium by means of one-step nanosecond laser-processing. Thin sheets were micro-patterned by using 500mW ultraviolet laser pulses and a direct laser-writing technique. The microstructure contained blind micro-holes which improved the interface between water, air and solid, and thereby enhanced wetting. The geometrical changes were related to chemical changes at the surface, which improved the degree of hydrophobicity. The laser-processed micro-holes exhibited near-superhydrophobic behaviour, with a static contact-angle of 148°.

Anodic oxidation and self-assembly processes were used[14] to prepare superhydrophobic aluminium alloy surfaces having a water-contact angle of 157.5° and a sliding-angle of 3°. These resulted from an hierarchical micro-nano structure and the assembly of low surface-energy fluorinated components upon it. The untreated alloy substrate was hydrophilic, with a contact-angle of about 97.9°. After modification, but without anodic oxidation, the surface was hydrophilic with a contact-angle of about 114.7°. Water droplets spread completely on surfaces which were modified only by anodic oxidation. The essentially zero contact-angle suggested that the anodized alloy surface was superhydrophilic. This was because the volume of the hydrophilic alumina and the roughness of the surface greatly increased during anodizing. A transition from superhydrophilicity to superhydrophobicity was possible by adjusting the modification process for the surface. The superhydrophobic surface maintained its nature following abrasion.

Superhydrophobic surfaces were produced on aluminium by shot-peening, dislocation-etching and immersion in solutions of stearic, myristic and decanoic acids[15]. Dislocation-etching produced an hierarchical structure at the nano-scale and micro-scale. Surfaces which were etched with stearic acid had a low surface energy, with a contact-angle of about 157° and low adhesion for 8μL drops. Increasing the etching-time increased the static contact-angle, because it increased the roughness and number of micropores. According to the Cassie-Baxter law, the hierarchical structure trapped air-bubbles and decreased the area of contact-surface with water.

Superhydrophobic surfaces with controllable adhesion were prepared by means of femtosecond laser ablation. The adhesion could be modified from being extremely low to high by adjusting the laser-processing parameters[16]. Various hierarchical structures with micro/nano-scale features could be produced by adjusting the processing parameters so as to produce a range of wetting abilities. Cleaned samples were irradiated with linearly-polarized light having a wavelength of 1030nm, using a repetition-rate of 75kHz and a pulse-width of 1000fs. The output-power ranged from 1000 to 10000mW. The resultant changes in surface morphology depended very much upon the scanning-speed, laser-power and scanning-interval and could cause a transition from Cassie to Wenzel behaviour.

Laser-texturing was used to create[17] a regular dimple-pattern on surfaces which, following stearic-acid treatment, became superhydrophobic. Five types of regular dimple-pattern arrays were produced on the surface, having diameters of 0.2, 0.4, 0.6, 0.8 or 1.0mm. The wettability of laser-textured samples could be regulated by choosing the dimple dimensions during laser-processing. When the diameters of the dimples were increased, the water contact-angle decreased and the water sliding-angle increased.

Foil with a roughened surface was prepared by anodic treatment in neutral aqueous solution[18], and pitting corrosion by NaCl. The surface of untreated foil was smooth, but became coarse, with a labyrinthine hole-and-step topology following anodization. The original hydrophobic surface, with a contact-angle of about 79°, became superhydrophilic, with a contact-angle of less than 5°, following anodization. The superhydrophilic surface then became superhydrophobic, with a contact-angle greater than 150°, following modification with oleic acid.

A superhydrophobic surface was created on an aluminium substrate by using a sol–gel method which involved immersing the clean pure substrate into a solution of zinc nitrate hexahydrate and hexamethylene tetra-amine[19]. Following heating (95C, 1.5h), it was modified with alkane thiols or stearic acid. When the molar ratio was changed from 10:1 to 1:1, the contact-angle was greater than 150°. The best surface had a water contact-angle of about 154.8°, with an angle-hysteresis of about 3°. The surface of the films comprised ZnO and Zn-Al layered double-hydroxide in a flower-shaped arrangement. This flower-like porous structure, and the low surface energy, led to superhydrophobicity.

A static water contact-angle of about 154° was obtained by depositing stearic acid onto an aluminium alloy, and the coating exhibited a circa 30° contact-angle hysteresis. Superhydrophobic surfaces with a static contact-angle of about 162° and 158°, and a contact-angle hysteresis of about 3° and 5°, respectively, were obtained by incorporating nanoparticles of SiO_2 and $CaCO_3$ into stearic acid. The high hydrophobicity was attributed to the synergistic effects of micro/nano-roughness and low surface energy. Study of the wettability between 20 and -10C showed that the superhydrophobic surface became hydrophobic at low temperatures.

Table 1. Effect of HCl-etching time on aluminium surfaces

HCl (min)	Modification	Contact-Angle (°)	Rolling/Sliding-Angle (°)
0	a	74.7	-
4	a	25.6	-
4	a	169.2	4.2
2	a	107.7	135.7
3	a	154.3	21.9
6	a	168.0	5.1
8	a	166.6	3.8
4	b	155.9	45.7
4	c	164.8	6.8

a) 10min modification with dodecanethiol-myristic mixture, b) 10min modification with dodecanethiol, c) 10min modification with myristic acid

Materials Research Forum LLC
https://doi.org/10.21741/9781644903896

An homogeneously structured superhydrophobic surface with a gradient non-wettability was created[20] by combining chemical etching and vapour-diffusion modification. The as-prepared surface exhibited a marked gradient of water repellence, with a water contact-angle of between 162 and 149°. The sliding-angle exhibited a corresponding change from 3 to 11°. The gradient nature of the non-wettability led to a droplet surface-adhesion that ranged from 19μN at the most hydrophobic end to 57μN at the other end. Because of the difference in water-adhesion force, droplets tended to roll well in a specific direction.

Table 2. Contact-angles of etched aluminium surfaces

Etchant	Advancing Contact-Angle (°)	Receding Contact-Angle (°)	Hysteresis (°)]
none	85	17	67
FeCl₃	155	149	7
CuCl₂	158	149	9

A directional caterpillar-like rolling of droplets along the ridges of inclined ratchet-like superhydrophobic surfaces has been observed[21]. The superhydrophobic coating comprised micron-scale particles and 50 to 100nm nano-holes. In the opposite direction, droplet movement depended only upon the end of triple-phase contact-lines while the front of the contact line was pinned. Sliding-angle measurements (table 1) indicated that the ratchet-like superhydrophobic surfaces exhibited a directional drop-retention behaviour. A reduction in the rise-angle, in the height of the ratchet's ridge and in the volume of the droplet could markedly increase the directional difference in droplet retention on the ratchet-like surfaces. It was concluded that superhydrophobicity and periodic ratchet-like microstructures were pivotal to directional droplet-sliding in the one-dimensional case.

Flexible superhydrophobic aluminium sheet was used to subject droplets to rolling and coalescence. This revealed the conditions which yielded the best superhydrophobic surfaces; those with the highest water contact-angles and lowest sliding-angles, as well as the shortest contact-time during droplet bouncing. The best water-repellent properties, as indicated by the contact- and rolling-angles (table 1), were obtained by HCl-etching. pristine aluminium was usually smooth at the micrometre scale, with just a few scratches and patch-like defects. Following HCl-etching, the water contact-angle decreased from about 76° to about 26°. This was attributed to a Wenzel-type contact, in which surface-roughness increased wettability.

Superhydrophobic aluminium surfaces have been investigated[22] as a means for encouraging drop-wise condensation. The required superhydrophobicity was obtained by etching, and then depositing fluorosilane so as to lower the surface energy. Various etchants nano-textured aluminium surfaces to give differing surface morphologies. The etchants led to contact-angles of about 156° and contact-angle hystereses of less than 10° at room temperature (table 2). Both substrates promoted drop-wise condensation. In a saturated-vapour environment, condensate droplets grew and moved on the surfaces in the Wenzel state.

Cyclic chemical etching was used[23] to create superhydrophobic surfaces with a honeycomb structure. Samples which were etched 8 times exhibited micro-nano scale honeycomb cavities and had a water contact-angle of 135°. When further treated with octadecane thiol methanol solution, the contact-angle was 153.1°. The contact-angle of the unetched surface was about 46.1°, making it hydrophilic due to the inability of the smooth surface to capture air. The angle increased rapidly with increasing etching time. The roughness increased from 117nm to 779nm with increasing etching time, and the change in surface morphology from smooth to honeycomb pores resulted in an increase in the amount of air which was captured. Modification with octadecane thiol covered the etched surface with a monomolecular film which was formed by adsorption and reaction between the octadecane thiol and the etched surface, thus reducing the surface free-energy and producing a superhydrophobic surface.

Aluminium surfaces were treated electrochemically, and the addition of hydrochloric acid to the electrolyte had an appreciable effect upon the wettability[24]. The static water contact-angles were 115.6° and 129.2° when the electrodeposition times were 5min and 10min, respectively. Surfaces which were prepared by using 2V for 20min had a static water contact-angle of 155.8° when 0.015M HCl was added. The static contact-angle remained above 150° following exposure to 3.5wt%NaCl solution for 30 days. The low contact-angles for short electrodeposition times were attributed to insufficient etching of the aluminium. Only scratches originating from polishing were present on plain samples, while the superhydrophobic surface was much rougher and was covered with leaf-like sheets of various sizes. Compact micro- and nano-sheets which were stuck to the surface could harbour considerable amounts of air and thus exhibit superhydrophobicity.

A superhydrophobic surface was prepared on an aluminium substrate, with anodization and low-temperature plasma treatment being used to create a micro–nano structure and with trichloro-octadecyl-silane being used to modify the roughened surface[25]. The resultant static water-contact angle was 152.1°. A rougher surface, having micro-nano pores and other features, was produced when a low-temperature plasma treatment was applied to anodized film, resulting in a static water-contact angle of up to 157.8°.

Nanoparticles of ZnO were deposited onto commercial-purity aluminium via simple immersion and ultrasound treatment[26], followed by surface-energy reduction using stearic acid in ethanol. Ultrasound led to more stable superhydrophobic aluminium surfaces than did simple immersion. Etching in hydrochloric acid could be used before ZnO deposition in order to create a more mechanically stable superhydrophobic surface. The distinctive topography could have a marked effect upon the wettability and stability of superhydrophobic surfaces. With increasing immersion time, the water contact-angle increased and reached a plateau after 0.25h. The sliding-angle fell to its lowest value at this point.

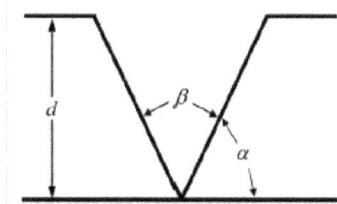

Figure 1. Geometry of laser-cut nanogrooves

Table 3. Average angles and depths of V-shaped microgrooves as a function of laser fluence

Fluence(J/cm^2)	β(°)	Depth(μm)
11.63	66.55	35.45
18.49	58.16	39.64
27.71	55.52	43.10
35.55	53.38	46.13
44.05	48.04	53.20
52.67	40.21	62.50
61.56	35.89	63.17
68.42	33.92	61.68

Arrays of parallel V-shaped microgrooves were created[27] on an aluminium surface, by means of femtosecond laser machining (figure 1, table 3), in order to obtain a super-wicking surface which could quickly transport water uphill. The relationships between flow-time and flow-distance were compared with theory. The laser fluence and scanning step-size affected the super-wicking behaviour. Surfaces which were fabricated by using a laser fluence of 18.49 and 52.67J/cm^2 offered the best super-wicking performance, with average water-flow velocities of about 16.2 and 16.4mm/s, respectively, over a distance of 30mm. During the same time, surfaces which were fabricated using 52.67J/cm^2 could transport a greater (approximately 1.5 times) volume of water than could a surface which was fabricated at 18.49J/cm^2. The super-wicking surfaces exhibited anisotropic flow due to the parallel microgrooves. When the scanning step-size was reduced to 25µm, the surface formed irregular rough structures which resulted in isotropic flow. The key to obtaining a high-performance superwicking surface is to process deeper and regular microgrooves.

Biomimetically-inspired by cactus spines and pitcher-plants, geometry-gradient slippery surfaces were made[28] from aluminium alloy which produced directional motion due to topographic gradients. The mechanism was related to a competition between a driving force which was produced by the Laplace pressure, and an adhesive force which was due to viscous resistance. The lateral adhesive force on the geometry-gradient slippery surface was much smaller than that on the original surface. The slippery surface was created by one-step nanosecond laser ablation and oil spin-coating. This led to the continuous self-transport of water and oil droplets on horizontal surfaces and uphill self-transport on the inclined surface. Experiment showed that the adhesive force between the original aluminium surface and a droplet was much greater than that on the geometry-gradient slippery surface. This prevented spreading of the liquid droplet on the original triangular surface. A droplet of large volume could travel a greater distance and a small vertex angle could result in a higher velocity. The mechanism of antigravity self-transport was carefully studied by constructing a theoretical model and noting the effect of droplet volume on the incline angle and distance. The mechanism was related mainly to a competition between driving force and resistance. The latter was made up of two main components. One was the adhesive force which acted along the incline and the other was the droplet-weight, which acted along the inclined. When the driving force which was induced by the gradient geometry, the Laplace-pressure difference, was greater than the sum of the adhesive force and weight, a droplet moved up an inclined geometry-gradient slippery surface. With increasing transport distance, the driving force decreased and the droplet finally stopped in a pinned state. Quantitative experiments explored the effect of the incline angle, the vertex angle of the geometry-gradient slippery surface and droplet

Materials Research Forum LLC
https://doi.org/10.21741/9781644903896

volume on antigravity self-transport. These showed that the travel distance decreased with increasing incline angle, and this was related mainly to the increase in resistance which was due to droplet weight acting along the incline. In addition, the transport distance increases with the decreasing vertex angle at the same climbing angle. The droplet-volume contributed to improving the travel distance, and the greatest distance was about 4mm when the droplet volume was 7μL. The critical incline angle as a function of droplet volume on geometry-gradient slippery surfaces with vertex angles of 15, 20, 25, 29, 41 and 53° was determined and ranged from some 28 to 63° for volumes ranging from 4 to 8μL. The critical incline angle was defined to be the greatest angle up which the droplet could self-transport. The critical angle was also related to the vertex angle, due possibly to the shape of the droplet when tilted.

Copper

Superhydrophobic surfaces on 2N5-purity copper were obtained[29] by combining etching in 10wt% ammonia solution with calcination at 340C. The surface was further modified by using an ethanol solution which contained stearic acid. The resultant surfaces had contact-angles as high as 157.6°. The untreated copper surface was very smooth, with a contact-angle of 76.5°. Following etching for 20h, there were irregular cell-like projections with a height of several micrometres on the surface. That surface had a contact-angle of 21°.

When a 30V direct-current voltage was applied between two copper plates, 1.5cm apart, immersed in dilute ethanolic stearic acid solution[30], the surface of the copper anode became superhydrophobic due to a reaction between the copper and the stearic acid, such that it became covered with flower-like low surface-energy copper stearate. This imparted a water contact-angle of 153°, and roll-off properties.

Wet chemical-reaction was used to create a superhydrophobic surface on a polished copper substrate at room temperature[31]. The surface had a water contact-angle of about 154° and a sliding-angle of about 4°. These were attributed to the roughening caused by the chemical reaction and to the low surface free-energy which was imparted by treatment with vinyl-terminated polydimethylsiloxane.

Etching and hydrothermal treatment were used[32] to produce a superhydrophobic surface on copper with a contact-angle was 157.7°. The bare copper surface had a contact-angle of 76.5°. A few microns of pebble-like structure appeared on the etched copper surface, and the contact-angle was only 21°, thus indicating that the copper surface was uniformly dissolved by the $NH_3 \cdot H_2O$ solution. Surfaces of 3N-purity copper were modified[33] using nanosecond laser and 1H,1H,2H,2H-perfluoro-octyltri-ethoxysilane or re-filled nano-silica. The inverted-pyramid microstructures were arranged in a continuous regular pattern, with the outer layer constituting a compressive wear-resistant surface and the inner layer being filled with hydrophobically-treated nano-silica. The contact-angle of the surface was 160.3° and the rolling-angle was 1°. When copper-based superhydrophobic materials were prepared by means of oxidation, lauryl mercaptan-modification and compression moulding[34] the surfaces had a cauliflower-like structure, with long lauryl mercaptan chains self-assembled onto them. The contact-angle was 155.2° and the sliding-angle was less than 5°.

A one-step electrodeposition method has been used to create a micro-nano superhydrophobic structure on the surface of copper by using a choline chloride based electrolyte[35]. As compared with bare copper, with a contact-angle of 62.2°, the contact-angle of the electrodeposited coating attained 157.8°.

Superhydrophobic 4N6-purity copper surfaces were prepared by means of oxidation involving $NaClO_2$, $NaOH$ and $Na_3PO_4 \cdot 12H_2O$, followed by modification with 1-octadecanethiol[36]. The optimum conditions were an oxidation time of 0.5h, a modifier concentration of 2.5mM and a modification time of 5h, leading to a contact-angle of 161.1° and a sliding-angle of 2.2°. When the surfaces were immersed in de-ionized water and exposed to air for several days, the contact-angle was reduced to 119° by immersion in water for 1 week, but was not affected by an air-exposure of 45 days.

A template and etching method was used to create a regular hierarchical multi-scale structure on foil by using the surface of bamboo leaf as a template[37]. This structure increased the water contact-angle of the foil surface from 64° to 131.1°. The hierarchical structure could then be further modified using stearic acid, leading to a contact-angle of 160.0° and a sliding-angle of 3°.

Chemical etching was used to produce superhydrophobic surfaces by initial immersion in ferric chloride solution[38]. The etched surfaces had a maximum contact-angle of 140°. They had a high sliding-angle, and water droplets were retained even on inverted surfaces. Following stearic acid modification of the etched surfaces, the contact-angle increased to more than 150° and the sliding-angle decreased to less than 10°.

Coatings with copper deposits were prepared by using jet-electrodeposition[39]. The coatings had a micro-nano structure. Following modification with stearic acid, a superhydrophobic surface was obtained. The static contact-angle and the sliding-angle of the surface were 151.6° and 5.7°, respectively.

A superhydrophobic copper surface was created by using electrochemical deposition and lauric acid functionalization[40]. The surface had contact-angles as high as 158°. Periodic surface ripples were produced on 3N-purity copper by using picosecond laser (1064nm wavelength, 203.6kHz repetition-rate, 10ps pulse-width) nanostructuring. Following modification with triethoxyoctylsilane, various types of ripple exhibited differing levels of wettability[41]. Fine ripples, with few re-deposited nano-particles, exhibited a high attraction to water. An increased amount of nano-scale structure decreased the adhesive force to water and also increased the contact-angle. One specific type of ripple exhibited superhydrophobicity, with a contact-angle of 153.9° and a sliding-angle of 11°.

Nanosecond laser-processing and sol-gel methods were used to produce micro-nano inverted-pyramid structures, modified with SiO_2-polydimethylsilane, on copper[42]. The superhydrophobic surface had a water contact-angle of 159.5° and a sliding-angle of 0.5°. The middle SiO_2-polydimethylsilane coating conserved superhydrophobicity when the top-most nanostructure was worn. As wear continued, the inverted-pyramid micro-nano structural array protected the internal nano-hydrophobic material. The adhesion of water to the surface was 3.27µN.

Superhydrophobic surfaces were produced on copper plate by treatment with $AgNO_3$ and dodecyl mercaptan[43]. The as-prepared surfaces had a hierarchical rough structure which comprised nano-sheets and nano-particles. Long alkyl chains were assembled on the rough surface. This led to a water contact-angle of 156.8° and a rolling-angle of about 3°. Its impressive performance was attributed to the so-called cushion-effect and to capillary phenomena.

The wetting-transition from the Cassie mode to the Wenzel mode degrades the performance of superhydrophobic surfaces. The wetting-transition occurs when the surface tension can no longer resist the gravitational force, and the liquid penetrates the spaces between asperities; leading to collapse. The wetting stability of 3N-purity copper-based superhydrophobic surfaces was investigated[44]. The samples had nano-asperities with a diameter of 70nm, but with two different packing-densities. The static (sessile-droplet) and dynamic (drop-wise condensation) wetting stabilities were compared. Sessile droplets on surfaces with densely-packed nano-asperities having a pitch of 120nm remained in the stable Cassie mode. The wetting-transition from Cassie mode to Wenzel mode occurred spontaneously on coarsely-packed nano-asperities having a pitch of 300nm. The contact-angle on the surfaces of coarsely-packed nano-asperities decreased from over 150° to about 110°, and the sliding-angle increased from less than 5° to over 60° within 200s. There were essentially no changes on the other surface. Condensed droplets on surfaces with densely-packed nano-asperities remained in stable Cassie mode, whereas the droplets on surfaces with coarsely-packed nano-asperities were in Wenzel mode.

Hierarchical micro-nano scale binary rough structures were created on copper by 30V direct-current electrochemical machining in a neutral 0.2mol/L NaCl electrolyte[45]. It required only 3s to produce the necessary roughness. The rough structures comprised micrometre-scale potato-like features and nanometre-scale cube-like features. Following modification with fluoro-alkylsilane, the surfaces were superhydrophobic, with a water contact-angle of 164.3° and a tilting-angle of less than 9°.

Table 4. Surface energy of milled copper

Cutter Tip Distance (mm)	Energy (mN/m)
-	47.296
25	0.324
30	0.259
35	0.225

Copper sheet was etched in 25wt% ammonia solution, under ultrasound, so as to produce a uniform roughness, and then immersed for 120h in a 0.02mol/L solution of octadecanoic acid in 35C ethanol before annealing (120C, 1h) to give a superhydrophobic coating[46]. The ultrasonic treatment shortened the etching-time and improved etching uniformity. The static contact-angle was up to 157°, and the contact-angle hysteresis was 4.2°. The proposed mechanism was that copper atoms reacted with octadecanoic-acid molecules and a long carbon-chain structure formed on the surface, imparting superhydrophobicity.

Milling, deposition of $AgNO_3$ solution and modification with stearic acid were used[47] to create superhydrophobic copper surfaces (table 4). The surface morphology was dendritic and rectangular surface promontories, produced by the milling, were distributed over the substrate. The water contact-angle could be as high as 158.4°. When scratched with a knife and abraded, the substrate retained good superhydrophobicity. The wettability of the surface after wear-tests showed that the contact-angle was 152.7°; lower than the value of 157.6° before friction tests. After further testing, the contact-angle remained above 148°. The surface obeyed the Cassie model, with a high contact-angle and low viscosity.

Another superhydrophobic surface was produced[48], having a water contact-angle 147° and a roll-off angle of 5°, was created by annealing foil in air and coating it with silica nanoparticles that were dispersed in a silane solution. There was a uniform outer distribution of spherical micron-sized CuO particles, and an inner layer comprising a mixture of CuO and Cu_2O. The silane-coated surfaces exhibited augmented valleys and peaks having a higher root-mean-square and average roughness, due to the silica nanoparticles. The critical surface energy of the superhydrophobic surface was calculated to be 17.72mN/m.

Table 5. Contact-angles of variously processed copper surfaces

Morphology	Fluorinated	Contact-Angle (°)	Nature
micro-nano rough	no	20.5	strongly hydrophilic
micro-rough	no	60.0	strongly hydrophilic
nano-rough	no	73.0	strongly hydrophilic
flat	no	78.0	hydrophilic
flat	yes	132.0	hydrophilic
nano-rough	yes	137.0	hydrophilic
micro-rough	yes	149.3	hydrophilic
micro-nano rough	yes	164.0	superhydrophilic

Sandblasting and hot water treatment were used to create micro-roughness and nano-roughness, respectively[49]. Bare sheets were sandblasted using Al_2O_3 particles, while hot-water treatment involved simple immersion in de-ionised water at 75C for 24h. The processing resulted in copper oxide nanostructures, formed by the hot water treatment, coated onto the microstructured surface which was produced by sandblasting. The nanostructures possessed CuO stoichiometry and took the form of leaves, with a thickness of 15nm and a width of 250nm. The nano-leaves, when located at the side-walls of micro-hills, were critical to imparting superhydrophobicity. The hierarchically rough samples were coated with 1H,1H,2H,2H-perfluorodecyltrichlorosilane in order to reduce the surface energy. The resultant water contact-angles were as high as 164°. Fluorinated, nano-rough and micro-rough surfaces exhibited other angles (table 5). Some non-fluorinated micro-nano rough surfaces exhibited a strongly hydrophilic behaviour, with a contact-angle of 50°.

Micro-scratches and grooves were visible on pristine surfaces. At the nano-scale, the surfaces appeared to be leaf-like. These sharp dense structures contributed to the high roughness which was associated with the substrates. The surface contained copper and oxygen, suggesting that the leaf-like nanostructures were CuO. The average surface roughness of the surfaces was 428nm. The average water contact-angle was 83° for the pristine copper surface, 110° for the copper surface when coated with trichlorosilane and 169° for the superhydrophobic surface.

Superhydrophobic copper surfaces with a contact-angle of 156.2° and a sliding-angle of 4° were prepared by means of hydrothermal treatment and silane modification[50]. The superhydrophobic Cu_2S-coated surface comprised a large number of Cu_2S crystals grafted to long hydrophobic alkyl chains. The material thus had a rough hierarchical surface with micro- and nano-scale features. Heat treatment at 200C had a marked effect upon the surface microstructure and wettability.

Superhydrophobic surfaces were prepared by oxidation, heat-treatment and alkyl-chain grafting[51]. The surfaces had a structure which comprised CuO nano-sheets and needle-like fibres. The micro-nano scale hierarchical surface and grafted long alkyl chains imbued the surface with water-repellence, and the water contact-angle and sliding-angle could attain 157.3° and 5°, respectively, after modification with stearic acid. The contact-angle gradually increased with increasing immersion time up to 24h. It then decreased with increasing immersion time. This was attributed to Cassie-Baxter behaviour. When treated in stearic acid ethanol solution for less than 24h, the alkyl chains grafted onto the surface. During continued immersion, more stearic acid molecules were grafted. More and more air was therefore trapped in the solid/liquid contact area under a water droplet. When the immersion time was too long, the number of grafted stearic acid molecules saturated. Some molecules were then deposited on the surface in the form of physical adsorption, causing the contact-angle to decrease.

Stable superhydrophobic surfaces were produced by using 1062nm laser-ablation to produce line and grid patterns[52]. Ablation increased the surface roughness, increased the number of (111) planes and decreased the number of (200) planes. The rapid evolution of the patterned surface from hydrophilic to superhydrophobic caused a rapid change from hydrophilic CuO to hydrophobic Cu_2O and organic adsorption. The wetting properties could be modified by changing the step size. The raw surface had a hydrophobic contact-angle of 119°, but it changed to hydrophilic following laser-ablation. The angle decreased from hydrophobic to hydrophilic due to the CuO structure on the surface. The contact-angles of the line and grid patterns generally increased upon increasing the step-size from 20 to 150μm; apart from the step-size of 150μm for the line pattern. The angle increased because of the raised non-ablated and hydrophobic surface. The angle decreased to 43° on a line pattern with a 150μm step-size, and this was attributed to the line patterns having an irregular ablation-spot distribution. The contact-angle increased with time such that, after 5 days, the surface changed from hydrophilic to hydrophobic for line patterns with 100μm steps or grid patterns with 20 or 100μm steps (table 6).

Table 6. Contact-angles of laser-patterned copper surfaces

Pattern	Step-Size (μm)	Condition	Contact-Angle(°)
line	20	ablated	16
line	20	aged, 5days	105
line	100	ablated	48
line	100	aged, 5days	155
line	150	ablated	43
line	150	aged, 5days	117
grid	20	ablated	20
grid	20	aged, 5days	152
grid	100	ablated	40
grid	100	aged, 5days	152
grid	150	ablated	57
grid	150	aged, 5days	111

When the step size of the line pattern and the grid pattern was 150μm, the contact-angle decreased to 117° and 111°, respectively. The decrease in angle with increasing step-size was attributed to easier droplet contact with flat areas. A line pattern with a step-size of 100μm had the highest contact-angle because that pattern had the highest height/diameter ratio, and the latter determined the effectiveness of air-pocket formation under the water surface. An increase in depth led to an increased angle because the droplet could not touch the bottom of the valley. When the depth was not too great, the droplet touched the bottom, and that promoted Wenzel behaviour at the surface. Following annealing (110C, 2.5h), the contact-angle increased from less than 70° to greater than 150° (table 7); apart from a line pattern with a 20μm step-size. The wettability in the latter case changed from hydrophilic to hydrophobic surface because air-pockets could occur only in the main groove of the ablation area.

Table 7. Contact-angles of laser-patterned copper surfaces

Pattern	Step-Size (μm)	Condition	Contact-Angle(°)
-	-	ablated	119
-	-	heat-treated	111
line	20	ablated	16
line	20	heat-treated	139
line	100	ablated	48
line	100	heat-treated	157
line	150	ablated	43
line	150	heat-treated	155
grid	20	ablated	20
grid	20	heat-treated	157
grid	100	ablated	40
grid	100	heat-heated	160
grid	150	ablated	57
grid	150	heat-treated	163

When the step-size became small, deformed copper covered the protuberant part and welded the island so that the surface then had a semi-regular pattern. The contact-angle on grid patterns was more stable because those patterns had a small portion of flat

Materials Research Forum LLC

https://doi.org/10.21741/9781644903896

surface, surrounded by considerable trapped air which prevented liquid from wetting the surface.

Laser-ablated textures exhibited a lower adhesion-force and sliding-angle than did those of the raw surface. A decreasing adhesion-force favoured a low sliding-angle (table 8).

Superhydrophobic surfaces with dense nanostructures were created on substrates by means of template-assisted electrochemical deposition[53]. During deposition, the bubbles which were generated persisted in some regions and prevented the creation of the nanostructure, thus producing an heterogeneous surface. In order to assure an homogeneous surface, the electrolyte composition and the voltage had to be optimized.

Table 8. Adhesion force and sliding-angle of laser-patterned copper surfaces

Pattern	Size (μm)	Condition	Adhesion Force (μN)	Sliding-Angle(°)
-	-	-	223.1	70
-	-	heat-treated	224.4	70
line	100	aged, 5 days	64.4	15
line	100	heat-treated	61	15
grid	100	aged, 5 days	59	4.3
grid	100	heat-treated	50	4.3

Droplet climbing which was driven only by surface-energy was found, and a superhydrophobic mesh was designed[54] which could pump water centimetres-high. An antigravity transportation system was moreover demonstrated that could continuously pump water droplets with no additional driving force. *In situ* monitoring was used to track the progress of a 5μL water droplet on a superhydrophobic substrate having low contact-angle hysteresis. When the droplet contacted the lower surface of the superhydrophobic so-called pump the droplet could then penetrate the mesh and merge into a water-column on the upper mesh surface, in the absence of any external force, leading to self-propelled rising of the liquid level in the tube. During penetration and ascent of the droplet, the 3-phase solid-liquid-gas contact-line did not extend horizontally. This differed from droplet-absorption by a hydrophilic substrate such as a sponge. When the droplet began to merge with the upper water-column, rapid release of its surface energy propelled antigravity delivery of the water. Wettability markedly affected the performance of as-prepared superhydrophobic pumps. A series of meshes having wettabilities ranging from superhydrophobic to superhydrophilic was used to investigate the effect of wettability on pump performance with regard to continuous water-droplet supply. As-prepared superhydrophobic mesh had a microscale/nanoscale wire roughness. The surfaces of hydrophobic and pristine meshes were relatively smooth. Droplets were provided by a hydrophobic syringe with an inner needle diameter of 100μm. The surface wettability of the mesh in the pump could affect the maximum self-driven height of the liquid level. When droplets with a radius of about 0.75mm and a volume of 1 to 2μL were used, hydrophobic meshes with a contact-angle of 150° and 137.5° could provide the greatest liquid-level rise: 13.6 and 10.3mm in height, respectively. When the liquid-level attained its greatest self-driven height, further addition of droplets to the lower end of the pump resulted in a lowering of the liquid-level in the tube. Water was then transported into the droplet and was rapidly spilt when the droplet and the upper water-column began to merge.

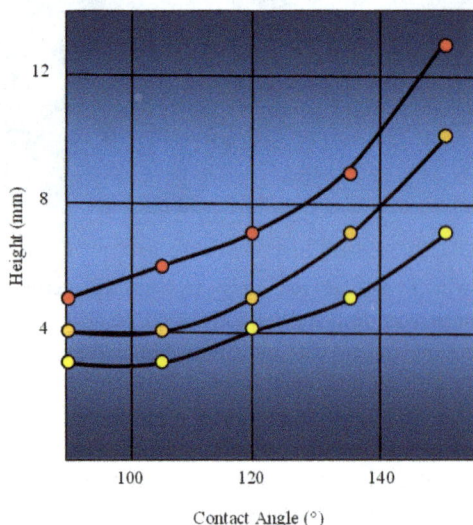

Figure 2. Maximum droplet ascent as a function of the water contact-angle on copper mesh. Red: 0.75mm, orange: 1.00mm, yellow: 1.50mm

When the liquid-level had collapsed, the original greatest self-propelled height could be reversibly restored by adding droplets. In the case of pristine mesh having a contact-angle of 115°, the height attained was only 2.5mm. Hydrophilic and superhydrophilic meshes could not provide water self-climb and no water-columns could be supported at their upper ends. The measured contact-angles were not entirely suitable for analysing the process, but the effect of wettability on the maximum self-driven climb of water (figure 2) was confirmed. The radius of the droplets appeared to influence the water-climb height of the pumps, together with the wettability of the mesh. According to the Young-Laplace equation, the upward Laplace pressure was inversely proportional to the water-droplet radius. An increase in the droplet radius resulted in a decrease in both the upward Laplace pressure and the greatest water-climb height. An integrated pump based upon superhydrophobic mesh with a pore-size of 500μm and a contact-angle of about 150° used droplets which ranged from about 0.75mm to about 1mm or about 1.5mm. The resultant self-climb height of water in the tube decreased from 14, to 12 and 7mm, respectively. Similar results were found for other hydrophobic meshes. In the limit, the droplet radius tended to infinity and a film could not be lifted by a superhydrophobic pump. It also resulted in collapse of the water-column above the mesh. In addition to the surface-wettability of the mesh and the size of the droplets, the effect of the mesh-size was investigated. The mesh-size markedly affected the initial state of the droplets. When the pore-size was less than 200μm, the droplets could not easily pass through the superhydrophobic mesh. When the pore-size was about 1mm, the superhydrophobic mesh could drive a climb height of less than 10mm.

A novel method, based upon a Laplace pressure imbalance, was exploited by means of track texturing[55]. This passively drove droplet motion, while maintaining the limited contact of the Cassie-Baxter state on superhydrophobic surfaces. The track resembled an out-of-plane backgammon-board with slowly-converging micro-ridges which were decorated with a nano-texture. A model was developed which explained the underlying mechanism. The rebound-angle of vertically landing droplets also led to horizontal self-transport to distances which were up to 65 times the droplet diameter and led to appreciable uphill motion. A copper surface was created which had hierarchical tracks upon which the droplets were propelled. The tracks resembled the sides of very acute and long isosceles triangles. Between each pair of tracks, a divergent microgroove was naturally formed. The combined action of the microgrooves, in contact with the liquid, led to an overall capillary-force imbalance which induced droplet motion on the tracks. In order to fabricate the microridges, laser-micromachining system was used. Due to the shape of the laser beam, a non-zero sidewall angle was created on the microridges, leading to a trapezoidal cross-section. The beneficial effect was that it generated an upward capillary force which aided the superhydrophobic behaviour of the droplet. In order to minimize adhesion forces and aid self-propulsion, the surface was made superhydrophobic by adding a second level of roughness to the tracks by using $Cu(OH)_2$ nanoneedles having a thickness of some 200nm. These clustered nanoneedles were then coated with a self-assembled monolayer of perfluorodecanenthiol. This reduced the surface energy of the substrate and made it superhydrophobic. The typical advancing and receding contact-angles of the nanostructured surface were 162.0° and 164.7°, respectively, leading to a contact-angle hysteresis of 2.7°. A droplet which was placed on such a surface formed elongated divergent menisci on its underside, with their number depending on the size of the droplet and the arrangement of the tracks. Each meniscus exhibited a linear dependence upon the radii of curvature at the rear and front. This produced a Laplace pressure-gradient along the meniscus. This gradient generated a propulsive force in the divergent direction of the microgroove between two consecutive tracks.

A bio-inspired superoleophobic pump was fabricated[56] from a superoleophobic mesh and an oil column. Depending upon the directional release of surface energy, oil droplets could be continuously collected, and pumped centimetres-high in the absence of any additional driving force. The so-called antigravity oil-delivery system could produce a continuous oil flow under water, and even air/water 2-phase oil transport. The copper-mesh device, with underwater antigravity oil transportation exhibited good reversibility. The directional oil-droplet permeability of the so-called superoleophobic pump in oil-water systems made it possible to collect oil droplets underwater. Because the mesh permitted oil-droplet penetration, droplets which were dispersed in water can be captured by the oil column. The superoleophobic pump could oil droplets from an oil/water mixture. It could lift the droplets when it was immersed in water, but the transport was interrupted by collapse of the oil column when it attained a critical height. The device, with its superoleophobic mesh, was also equipped with a bent quartz tube which permitted the continuous transportation of the oil droplets underwater. On the lower surface of the mesh, spherical oil droplets were continually furnished by syringe. Upon the release of surface energy, the droplets penetrated the mesh and, one after another, spontaneously joined the oil column, thus raising the level of the liquid. An oil droplet thus moved itself from a low position to a high position in the absence of any energy supply. In order to guarantee that the device was able to support the pressure from the oil column, underwater superoleophobic copper mesh with a pore size of 900μm was used. Oil droplets underwater were pumped to above the water level by some centimetres.

Water droplets on a heated surface are well known to be able to levitate over a film of evaporated water. When the surface has a ratchet-like saw-tooth topography, the droplets can even propel themselves uphill[57]. The extent to which droplets can be controlled is limited by the physical details of the Leidenfrost effect. Transition boiling can be induced, even at very high surface temperatures, and thus provide additional control of the droplets. Ratchets which have acute protrusions permit droplets to climb steeper inclines. Ratchets having sub-structures permit the direction of motion to be controlled by varying the temperature of the surface. On rough surfaces, the Leidenfrost effect acted when a droplet levitated to a sufficient height to leave no direct contact between the droplet and the most prominent features of the surface. Surfaces having sharper and more prominent ratchet-teeth were therefore expected to increase the Leidenfrost temperature. That then permitted droplets to be propelled more powerfully and to be able to climb steeper inclines. Three ratchets with teeth of differing sharpness were prepared by milling the brass surfaces using rotating blades. Block-1 had a pitch of 1mm, with the sloping part of the ratchet having an incline of 10°. Block-2 had a 1mm pitch, but with the sloping parts of the teeth at 30°. Block-3 had acute but asymmetrical teeth, with a pitch of 0.24mm. The blocks were heated, and spherical droplets of water with a diameter of 3.6mm were dropped onto them. The inclination of the block was increased until the droplets were no longer able to climb (figure 3). At the highest temperatures all of the blocks exhibited a similar trend, with the critical angle decreasing with increasing temperature. That is, increasing the temperature reduced the ability of the droplets to climb. This was typical of the Leidenfrost limit where, at higher temperatures, droplets are forced further away from the surface. With decreasing temperature the critical incline peaks before falling to zero. Block-3, with its sharpest teeth, exhibited highest peak-value of critical angle and the peak occurred at a considerable higher temperature than for the other blocks. This was attributed to be due to differences in the Leidenfrost temperatures of the surfaces. When a droplet made contact with a ratchet surface, the mechanisms which governed the resultant motion were complicated. It was not possible to neglect the effects of surface tension, and forces which were related to wetting contact with the surface. There was expected to be strong nucleate boiling and vibrations. In the case of block-3, where the critical incline exhibited a marked spike, contact with the surface could permit the droplets to grip the surface via wetting. Propulsion could meanwhile be maintained by the same gas flow as in the Leidenfrost regime. The droplets made contact with the saw-tooth peaks and were suspended between them, permitting film boiling in the suspended sections. It was anticipated that the adhesion which was associated with wetting would be too strong to be overcome by gas flow arising from film boiling. Vibrations which were caused by nucleate boiling could however provide enough activation energy to keep the droplets mobile. The droplets were opaque when the ratchet surface temperature was low, with violent movement of the droplet surface. At higher temperatures the droplets became smoother and more transparent. It was supposed that there was a strong effect of the greater gas flow at lower temperature which was associated with greater heat transfer to the droplet. This also increased the ability of the droplet to climb steeper slopes. In the case of block-2, it was difficult to decide whether a smaller sub-structure exerted an appreciable effect upon droplet motion. Droplet transport to the right was attributed to a complicated interplay between wetting, vibration and gas flow. Overall, additional control of droplets could be obtained by introducing greater contact between surfaces and droplets because this increased the Leidenfrost point, thus influencing droplet-dynamics and the boiling transition.

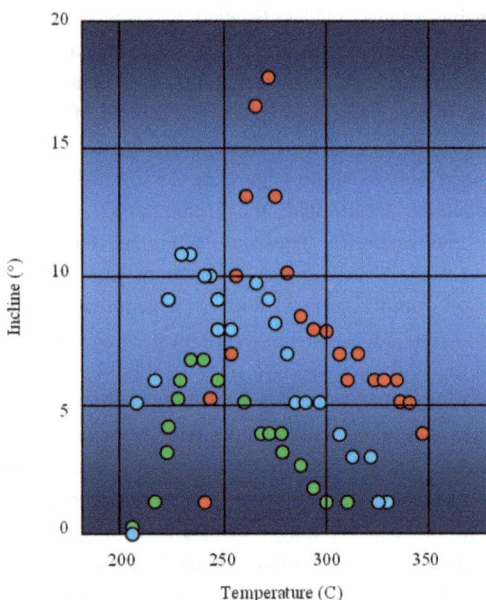

Figure 3. Maximum inclines which water droplets could climb on brass ratchet surfaces. Green: block-1, blue: block-2, red: block-3

Iron

Hierarchical structures were prepared on iron surfaces by chemical etching with hydrochloric acid, or by galvanic treatment in silver nitrate solution, and then modified using stearic acid[58]. The latter chemically bonded to the iron surface. As the HCl concentration was increased from 4 to 8mol/L, the surface became rougher and the water contact-angle changed from 127° to 152°. The nitrate concentration had little effect upon the wetting behavior, but a high concentration caused silver particle aggregates to change from flower-like to dendritic, due to the preferential growth direction of the silver. When compared with etching, the galvanic replacement method was more favourable to producing the roughness required for superhydrophobicity.

Superhydrophobic surfaces were prepared on iron by means of heat-treatment and surface-energy modification[59]. Micron-level structures were first constructed on the surface by scribing, followed by thermal oxidation to generate nanostructures upon the micron structures. The contact-angle was increased from 9.6° to 89.8° or 128.0° by scribing in various ways. The oxidation treatment changed the contact-angle to 21.7°, 122.2° and 138.1° for ordinary iron, iron which was scribed horizontally and vertically scribed and surfaces which were scribed multi-directionally, respectively. Following modification using silanes, the contact-angles were 107°, 131.1° and 157.3°. The greatest water contact-angle was 162.3°, together with a sliding-angle of 2.4° after annealing under the optimum conditions (450C, 4h). When the annealing temperature was 500C, the surface was unstable and the oxide layer peeled off. The process was applicable to surfaces which contained over 90% of iron. When the surface was too hard and wear-resistant it was difficult to carry out the required surface-roughening.

A process based upon K_2CO_3 was used to create 3-dimensional flower-like Fe_3O_4 micro-nano flakes on the surface of iron via *in situ* hydrothermal synthesis[60]. The width of the nano-flakes ranged from 50 to 100nm, with a length of about 3μm. The morphology of the Fe_3O_4 nanostructures could be varied from simple isolated nano-flakes to ordered 3-dimensional flower-like shapes by increasing the reaction temperature. The wettability of surfaces with the latter flower-like micro-nano flakes was changed from hydrophilic to superhydrophobic by modification with vinyl tri-ethoxysilane. The static water contact-angles on the modified surfaces were greater than 150°, and this was attributed to the modification and to the hierarchical structure. The surfaces retained a good superhydrophobic stability during long-term storage. When the reaction temperature was 120C, few flower-like micro-nano structures were formed. The surface had a contact-angle of 23° and a roughness of 114.2nm. When the temperature was increased to 180C, the flower-like structures grew larger, contiguous and essentially covered the surface.

The water contact-angle and roughness increased to 37° and 147.5nm. The increase from 23° to 37° was far from achieving the superhydrophobicity level. Following vinyl tri-ethoxysilane modification, the roughnesses of the surfaces obtained by 120C and 180C treatment were up to 155.1nm and 175.6nm, respectively, and the contact-angle was changed from 124° to 157°; indicating a change from hydrophobicity to superhydrophobicity. Surfaces which were obtained by 180C treatment had a sliding-angle of 1°.

Hierarchical structures were created on 2N5-purity iron by abrasion, calcination and modification, and the results showed that superhydrophobicity and adhesion depend upon micro-nano scale surface patterns[61]. The degree of adhesion could be controlled by adjusting the abrasion process so as to obtain patterns having an optimum ratio of height to width. Modification was carried out via immersion or vaporization, with the former being used to incorporate 1H,1H,2H,2H-perfluorodecyltri-ethoxysilane, octadecane thiol, oleic acid or sodium dodecyl sulphate and the latter being used to incorporate paraffin or lard. The resultant surfaces could be classified into superhydrophobic with high adhesion, superhydrophobic with moderate adhesion and superhydrophobic with low adhesion. The nano-scale patterns of high-adhesion and low-adhesion surfaces were not significantly different, but the micro-scale pattern of the former changed markedly to a relatively flat shallow structure with widths and depths of about 1μm. Typical values were a contact-angle of 158.3° and a hysteresis of 91.0°. The maximum adhesive force was of the order of 195.8μN. The pH-value had essentially no effect upon water-wettability or adhesion in the case of low-adhesion surfaces but, in the case of high-adhesion surfaces, pH-values of 2 and 1 led to contact-angles of 152° and 136°, respectively. The rolling-angle meanwhile decreased from 180° to 85° and 33°, respectively. It was surmised that high acidity destroyed the microstructure, given that the high-adhesion surface had no air-film in its pattern which could prevent direct contact between acid and surface. Between -10C and 200C, the low-adhesion surfaces maintained their contact-angle at 157° to 164° over the entire temperature range, while the rolling-angle decreased from 18° to 5° as the temperature increased from -10C to 30C. It then gradually decreased to 3° with increasing temperature. In the case of high-adhesion surfaces, the contact-angle increased from 150° at -10C, to 155° between 15C and 45C, and then decreased to less 150° at 70C. The rolling-angle remained at 180°, from -10C to 65C, and then decreased to 90° at 70C. This change was attributed to a transition from Wenzel behaviour to Cassie behaviour at higher temperatures due to the existence of a vapour film within the hierarchical structure which diminished contact with the rough structure. When exposed to atmospheric air, the low-adhesion surface retained superhydrophobicity for up to 120 days. The high-adhesion surface lost its superhydrophobicity after 80 days, but could be recovered. When immersed in 1M NaCl solution, the low-adhesion surface retained its superhydrophobicity for 5 days, while the high-adhesion surface could do so for only 3 days.

Leaves of *lotus, oxalis corniculata L., liriodendron tulipifera* and *fijoa sellowiana berg* were studied[62]. Studies of the static and dynamic wettability of the leaves revealed that superhydrophobicity was largely independent of the shape of nanoscale asperities and was mainly affected by the nanostructure size-distribution. This was exploited when creating biomimetic superhydrophobic surfaces having two scales of roughness. The behaviour of water droplets on superhydrophobic surfaces was investigated, showing that the surface nanostructure had a marked effect upon a droplet's dynamic behaviour. Biomimetic structures were constructed by depositing superhydrophobic films, using microwave plasma enhanced chemical vapour deposition. The superhydrophobic surfaces were created by covering roughened SiO_x films with self-assembled monolayers. The substrates were then electrostatically charged. Water droplets which fell on tilted charged surfaces jumped up, or ran uphill at high speed. This suggested that both the surface nanostructure and its chemical end-groups exerted an appreciable effect upon the motion of the water droplets. The substrates used were metallic plates and meshes made from stainless steel, or were glass slides. The surface morphology was examined using atomic force microscopy. The nanostructures minimized the effective contact surface and possibly increased the pressure in the air bubbles trapped below the water droplets. As a further investigation of the relationship between the 2-scale roughness and the dynamic behaviour of droplets, the surface was impacted with water droplets having a radius of 3mm which were dropped from a height. Complete bouncing occurred in the case of *lotus, oxalis* and *liriodendron* leaves. In the case of *feijoa* leaves, the droplet did not rebound. The rebound height decreased in the order: *lotus > oxalis > liriodendron*. The relationship between nanoscale roughness and energy dissipation was thus confirmed by the dynamic behaviour of the droplets, and a dual-scale roughness was advantageous for obtaining a superhydrophobic effect under many diverse conditions. By imparting 2-scale or multi-scale roughness, a surface could be optimized with regard to hydrophobicity. In dual scale roughness, it was the existence of the finer structure at the nanoscale which more strongly promoted hydrophobicity. The biomimetic structures had both a mm-scale roughness and a nanoscale roughness of 30 to 80nm. Due to the surface roughness, air was trapped below a drop and led to static contact angles close to 180° and a low dynamic contact-angle hysteresis. In the case of non-superhydrophobic surfaces, falling droplets immediately spread out as a result of the impact, and then shrank or receded so as assume a final shape. In order to move from one equilibrium position to another on a charged surface, the interface between the droplet and the surface had to overcome an

Materials Research Forum LLC
https://doi.org/10.21741/9781644903896

energy barrier. In a typical study, water-droplet motion occurred on superhydrophobic thin films which were created by coating glass substrates with -CF₃ or -CH₃ groups. Static charging was achieved by rubbing the underside of the substrate. The superhydrophobic films were prepared on previously roughened glass slides via thermal chemical vapour deposition of octadecyltrimethoxysilane or fluoroalkylsilane. Following charging, a water droplet was placed near to a surface which was tilted at a slope of 17° and the droplet jumped up, or ran uphill, on a -CF₃ coated surface. In the case of a -CH₃ coated surface, the droplet moved only slightly uphill. The higher build-up of static electricity was attributed to a difference in the surface potentials, with a CF₃-terminated surface possessing a 170 to 180mV lower potential than that of a CH₃-terminated surface. The droplet velocity was non-uniform, but was initially almost 500mm/s. When surfaces were uncharged, there was no running uphill.

An eccentric ring structure was developed which mimicked the directional transport ability of certain insect-parts, with stainless-steel substrates being subjected to femtosecond laser ablation in order to create micro/nano hierarchical textures[63]. Subsequent selective tungsten-sputtering produced superhydrophobic and superhydrophilic regions. This surface-energy difference promoted rapid low-resistance liquid movement along predefined channels. Water droplets fully wetted a 35mm eccentric-ring channel within 4.7s. The bio-inspired eccentric ring structure reduced resistance during water-transport, while continuous channels permitted rapid, long-range water diffusion. Modifying channel configurations permitted water to diffuse in various directions. The uniform distribution of surface energy on the material ensured homogeneous water movement along channels, and the low resistance led to a consistent transport efficiency, regardless of variations in the directions and overall channel distribution. Droplets moved preferentially along paths of lower resistance and higher capillary driving force, leading to asymmetrical wetting in various directions. This permitted directional prioritization without external control. The capillary force which acted on droplets on continuous tracks was sufficient to produce continuous upward movement, with 10μL of water being conveyed up slopes of 35 and 30°. On a slope of 35°, water flowed upward, but with some accumulation at the base. When the slope was 30°, water ascended the channel and fully wetted the surface.

Silver

A novel method was described for the spontaneous acceleration of droplets of liquid eutectic gallium-indium alloy to extremely high velocities through a liquid medium and along a predefined metallic path.[64] The droplet wetted a thin metal track (a 1mm-wide 100nm-thick film) and generated a force which both delaminated the track from the substrate, via spontaneous electrochemical reactions, and accelerated the droplet along the track. The familiar formation of a surface oxide on the eutectic prevented it from moving, but the presence of an acidic medium or the application of a reducing bias to the track constantly removed the oxide and permitted its movement. The track provided a sacrificial pathway for the metal. The liquid metal could accelerate along linear, curved and U-shaped tracks and travel uphill on inclines of 30°. The droplets could accelerate through a viscous medium at up to 180mm/sec; almost twice that of self-running liquid-metal droplets. In a somewhat humorous aside, it was pointed out that such a velocity was much higher than those of the fastest aqueous creatures when expressed in units of body-lengths per second. The eutectic of gallium and indium comprises 75wt% of gallium and is liquid at room temperature, given its melting-point of 15C. It possesses a low viscosity, a high surface tension and a high conductivity. It easily acquires a thin passivating skin of gallium oxide and this permits the alloy to adopt non-spherical shapes. In order to persuade the droplets to accelerate, advantage was taken of the high affinity of the alloy for other metals. Liquid gallium wets, and eventually dissolves, thin silver films. The wetting behaviour could be exploited if the leading edge of a droplet of the liquid metal wetted a silver film and pulled the droplet towards the rest of the silver. As the droplet moved, it would then dissolve the underlying silver and cause the rear edge of the droplet to adopt a high (non-wetting) contact angle. The resultant asymmetry would then propel the droplet. The alloy droplets unfortunately do not actually behave like that, and tend to be very unpredictable, hence the use of a track. When droplets were placed on thin metal tracks which were deposited on glass slides, the droplets ran only in the absence of surface oxide. The droplet delaminated and consumed the film as it ran. In spite of the high density of the eutectic, the force which acted on the droplet was sufficient enough to pull the droplet rapidly uphill on inclines, even through viscous 2M HCl (figure 4). The ratio of the droplet-diameter to the film-width, and the amount of track consumed by the liquid, affected the droplet-velocity.

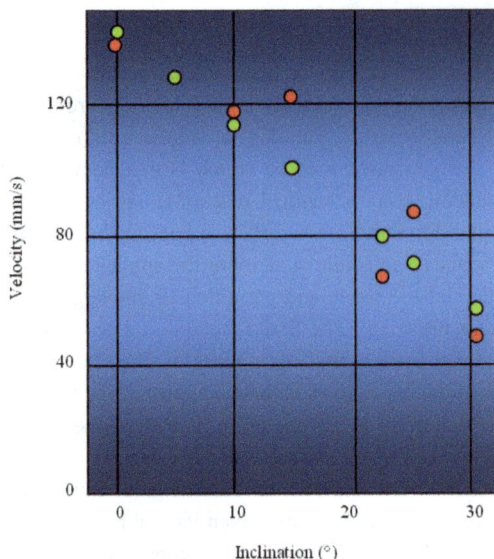

Figure 4. Measured and predicted velocities of 0.7mm-diameter gallium-indium eutectic alloy droplets on 1mm-wide silver/gold tracks at various angles. Red: measured, green: predicted

Titanium

Superhydrophobic surfaces were prepared by single-step anodization in an organic electrolyte, followed by stearic acid modification[65]. The material was anodized so as to produce a micro-nano hierarchical morphology which was then treated with stearic acid. The morphology of the surfaces could be modified by varying the anodizing time from 1 to 5h and the voltage from 5 to 50V. The static water contact-angle and contact-angle hysteresis on the best superhydrophobic surface (anodized for 3h at 35V) were 160.1° and 7°, respectively.

Superamphiphobic surfaces could be prepared by means of 1-step anodization and then fluoroalkylsilane modification. The prepared surfaces had water, glycerol and hexadecane contact-angles of 166.4°, 158.4° and 152.5°, respectively[66]. The corresponding sliding-angles were all within 10°. The figures were attributed to re-entrant micro-nano structures and the low surface-energy modification.

Superhydrophobic surfaces were prepared by anodization in sodium chloride solution, followed by immersion in perfluorodecyltriethoxysilane[67]. The 50nm anodic film comprised TiO_2 and $TiCl_3$ and had an hierarchical structure which consisted of a micro-scale horn structure and a nano-scale overlay. This surface had a water contact-angle of 151.9° and a sliding-angle of 3° following immersion. The surface coverage of the hierarchical structure was improved by mechanical attrition, which grain-refined the titanium. The thickness of 200nm, of the anodic film on the mechanically-treated surface, was clearly greater than the 50nm of the titanium surface. This was due to the large number of grain boundaries on the surface, which acted as rapid-diffusion paths during anodization. On the other hand, the adhesion of the mechanically-treated and anodized film was inferior to the film which was formed by anodization alone. This was attributed to a large number of pores within the former films.

A two-step process was used to create superhydrophobic surfaces on by shot-peening, followed by chemical etching[68]. The etched surface was then modified using methyltrichlorosilane in order to lower the surface energy. A nano-scale fibrous network was present on the surface, with chemical bonds existing between the functional groups of the methyltrichlorosilane and the titanium surface. The pre shot-peened surfaces had a maximum water contact-angle of 159° following methyltrichlorosilane modification.

Superhydrophobic surfaces with complex micro-pore structures and low surface roughness were created via anodic oxidation in $NaOH-H_2O_2$ solution. Fluoroalkylsilane was then used to reduce the surface energy of the electrochemically oxidized surface[69]. The as-prepared surfaces had a roughness of 0.669µm, with a water contact-angle of 158.5° and a tilting-angle of 5.3°.

Superhydrophobic surfaces were prepared on 4N-purity titanium by means of anodization and surface-energy modification[70]. The surface had rough micro-protrusions with a nano-flake morphology which resembled pine cones[71]. This microstructure, and the wettability of the surface, could be adjusted by altering anodization parameters such as the anodization time and voltage. The highest water contact-angle was 161.4° together with a sliding-angle which was essentially zero.

Table 9. Contact-angles on titanium for various
AgNO₃ immersion times

Time (h)	Contact-Angle (°)
0.5	144.7
2	151
5	152.7
7	154
12	152.3

It was noted that surfaces without chemical modification were superhydrophilic, but became superhydrophobic during exposure to air for long periods. Following high-temperature annealing, the time required for the superhydrophilic to superhydrophobic wettability transition, with a maximum water contact-angle of 153° and sliding-angle of essentially 0°, decreased from 81 days to 63 days. The wettability transition was due mainly to the adsorption of organic compounds from the ambient atmosphere and to air trapped in the microstructure.

Pulsed nanosecond laser ablation was applied to titanium alloy, combined with functionalization using polysilazane, so as to obtain a biomimetic lotus-leaf superhydrophobic surface[72]. The as-synthesized surface had a water contact-angle of 164.1° and a sliding-angle of 1.5°.

Hierarchical binary surface structures were obtained by hydrothermal treatment with successive solutions of oxalic acid and sodium hydroxide[73]. The hierarchical surfaces, following fluoroalkylsilane modification, had a maximum contact-angle of 158.7° and a sliding-angle of 4.3°. This led to an efficient self-cleaning behaviour, with bouncing and rolling-off of water droplets from the surface.

Superhydrophobic surfaces were prepared on 2N8-purity titanium which was pre-treated by mechanical polishing and anodizing, or by mechanical polishing alone. This was combined with the self-assembly of polydopamine and silver nano-particles, and post-modification using 1H,1H,2H,2H-perfluorodecanethiol[74]. The anodizing process could in fact be eliminated. The hydrophobicity increased with increasing deposition-time in silver nitrate solution (table 9). Surfaces which were so treated for 7h exhibited the optimum hydrophobic effect, with a water contact-angle of up to 154°. The surface was quite rough and was covered by relatively uniform micro-nano silver structures. The good hydrophobicity was attributed to the rough hierarchical microstructure, together with a low surface energy. The polarization curves of samples with and without a superhydrophobic surface were similar.

A micro-nano scale quasi-periodic self-organized structure was produced on surfaces, using a femtosecond laser ablation, which mimicked the surface of *Nelumbo nucifera* (lotus) leaves[75]. One scale consisted of large grain-like convex features which were between 10 and 20µm in size. The other feature, on the surface of the grains, comprised 200nm-wide irregular undulations. The use of these biomimetic surface patterns markedly changed the wettability of the surface. The original surface had a water contact-angle of 73°, but the laser-treated surface became superhydrophobic with a contact-angle of 166°.

Superhydrophobic surfaces were produced by using an ethanolic solution of myristic acid and hydrochloric acid plus simultaneous anodization and adsorption[76]. Using the optimum anodization potential, the surface was densely populated with hierarchical micro-nano clusters of titanium dioxide with adsorbed myristic acid. The maximum water contact-angle was 176.3°, with a sliding-angle of 1°. The surfaces had layered ridges which were decorated with sub-micron aggregates. The liquid/air areal fraction on the prepared superhydrophobic surface was about 0.97, with an asperity-slope greater than 71°. Under a 15V potential, the modified surface had a slightly increased contact-angle of 110.8°, as compared to the 67.6° of the bare specimen. When the potential was increased to 20V, the surface was now superhydrophobic with a contact-angle of 151.3°. Specimens which were anodized at 30V and 40V had contact-angles of 160.6° and 169.2°, respectively. The maximum angle was 174.4°.

Superhydrophobic surfaces were produced on 3N-purity material by using rapid-breakdown anodization, combined with stearic acid[77]. The anodized surfaces exhibited islands of TiO_2 micro-clusters, with a complex hierarchical structure, which were randomly distributed in a passive TiO_2 matrix[78]. Surfaces which were anodized at 30V were hydrophobic, with a water contact-angle of about 130° and a sliding-angle greater than 30°. Surfaces which were anodized at 50V were superhydrophobic, with a contact-angle of about 154° and a sliding-angle of 10° after stearic acid modification. When using molten stearic acid rather than a stearic acid-ethanol solution, the maximum water contact-angle was 167.8°, with a sliding-angle of 6°, on surfaces which had been anodized at 50V for 10min following molten stearic acid immersion at 125C for 0.5h. The same TiO_2 micro-clusters, with complex hierarchical structures and nano-pores were observed, and the micro-clusters were shown to be polycrystalline. Prolonged anodization promoted micro-cluster growth in each of their dimensions. The surfaces also exhibited vertical micro-plate crystals of stearic acid on the micro-clusters, and this increased the contact-angle.

Stable superhydrophobic surfaces were produced by chemical etching[79]. The best examples had a water contact-angle of 164° together with a water tilting-angle of about 2°. Superhydrophobic surfaces were prepared on Ti-6Al-4V by means of high-speed micro-milling, anodic oxidation and fluoroalkylsilane modification[80]. Regular microgrooves were constructed by micro-milling, and nano-tube arrays were created via anodic oxidation. Fluoroalkylsilane was then used to self-assemble a monolayer on the surface which had a micro-nano structure. Unlike the polished surfaces, the modified samples were superhydrophobic, with a water contact-angle of 153.7° and a contact-angle hysteresis of 2.1° (advancing and receding contact-angles of 153.8° and 151.7°, respectively). The surface of the polished material was smooth, while that of the machined material exhibited high and low staggered peaks and valleys, with traces of the milling at the bottom of the grooves and some processing burrs on the surface. Following anodic oxidation, closely-packed nano-tubes were present on the surface of the machined material, having an inner diameter of 25 to 30nm. They did not affect the structures produced by the micro-milling. Following fluorination, the surface morphology was not greatly changed.

A 1064nm pulsed picosecond-laser was used to create a micro-nano hierarchical structure on Ti-6Al-4V. The initial contact-angle of the polished surface was 71.6°. The imposed structure comprised dimple arrays having various diameters, depths and areal densities which were produced by controlling the pulse-energy and the number of pulses[81]. The contact-angle of the laser-textured surface was less than 30° and all of the surfaces were highly hydrophilic. The pitch between micro-dimples was 100, 80 or 60µm, corresponding to areal densities of 13, 20 or 35%, respectively. The aspect-ratio was 50%. The nano-features consisted of so-called laser-induced periodic surface structures, the dimensions of which could be varied by changing the laser energy-density and scanning-speed. The ripples had a period of about 1100nm when the energy-densities and scanning-speeds were 0.107 to 0.218J/cm^2 and 30 to 50mm/s. The contact-angle increased as the density of micro-textured surfaces increased. There was a slow (up to 4 weeks) transition from hydrophilic to hydrophobic. The wetting of the textured surfaces could be described in terms of the Cassie model. Following low-temperature annealing, the slow surface wettability transition could be markedly accelerated. This was attributed to changes in hydroxyl groups on the surface. The contact-angle of smooth surfaces did not change following the annealing treatment, indicating that it had little effect upon the wettability of the smooth surface. The contact-angle of textured surfaces was markedly changed. The contact-angle of surfaces with micro-dimples increased from 78.49° to

Materials Research Forum LLC

https://doi.org/10.21741/9781644903896

107.52°. The contact-angle of the rippled surfaces increased from 87.08° to 106.53° The angle increased from 101.27° to 136.79° when surfaces with micro-dimples were further covered with ripples. When the dimple surfaces were covered with periodic ripples, superhydrophobic surfaces with a contact-angle of up to 144.58° could be obtained.

Nanostructured superhydrophobic surfaces on Ti-6Al-4V were prepared by means of laser and anodizing treatments. The laser treatment generated a rough surface with parallel grooves and protrusions, offering superhydrophobicity following organic modification[82]. The anodizing treatment created a titanium dioxide nano-tube film. The contact-angle of the untreated alloy surface was 72°; that is, hydrophilic. Following the laser treatment, the angle decreased to 0°; that is, a superhydrophilic surface. Following the anodizing treatment, the angle remained equal to 0°. The Wenzel model explained the phenomena, in that the hydrophilicity of the hydrophilic surface increased with increasing surface roughness. The laser treatment produced a rough alloy surface, thus greatly increasing the surface roughness and reducing the contact-angle. The use of fluorosilane modification reduced the surface free-energy and provided superhydrophobicity. Following modification, the contact-angle was 103° for untreated material, thus transforming the original hydrophilic surface into a hydrophobic surface. Laser-treatment produced superhydrophobic surfaces with a contact-angle of 154°, and anodizing further increased the angle to 158°. The contact-angle of the superhydrophobic surfaces hardly decreased after 12h, and this was attributed to the high water-repellence of the surface. The laser treatment turned the morphology into one which consisted of peaks and valleys. The uneven structure could accommodate wear debris, reduce abrasive wear and improve the mechanical stability of surface features. A microstructure with evenly distributed peaks also offered a better mechanical performance and maintained superhydrophobicity. Anodizing further improved the mechanical stability of the superhydrophobicity. The nano-tube structure itself was difficult to remove, and the unworn nano-tube structure could also provide an extra air-cushion effect to aid superhydrophobicity.

Superhydrophobic lead surfaces were produced on Ti-6Al-4V substrates by 30 to 60s of immersion, followed by modification using low surface-energy materials. Hierarchical micro-nano scale structures were generated by immersion in $Pb(CH_3COO)_2$ solution, and were then modified using fluoroalkylsilane in order to reduce the surface energy[83]. The coatings had a water contact-angle of 165.5° and a sliding-angle of 4.6°.

Films of titanium dioxide with various nanostructures were produced on Ti-6Al-4V and commercial-purity titanium by means of electrochemical anodization[84]. The oxide films prepared on commercial-purity surfaces had a porous structure, while those on the alloy consisted of nano-rods. The oxide nano-film on the commercial material were hydrophilic, with a contact-angle of about 19°. The films on the alloy were superhydrophilic, with a contact-angle of less than 2°. Following treatment with fluorinated silane, the self-assembled films were now superhydrophobic, with contact-angles of 150° and 158°, respectively. The nanostructures and fluoroalkysilane were thought to play important roles in controlling wettability. Surfaces with nano-roughness alone did not exhibit superhydrophobicity.

A study of the wetting by liquid droplets of a Ti-6Al-4V micro-nano hierarchical hydrophobic surface clarified the transition from Wenzel to Cassie behaviour. Theory and experiment showed that a 1-dimensional nano-wire structure, which was produced on the surface by a hydrothermal treatment, changed the wetting-state of liquid droplets from Wenzel to Cassie due to its close dimensional correspondence to the micro-scale structure[85]. This increased the apparent contact-angle of liquid droplets on the solid surface to 161°, and also greatly decreased the sliding-angle, to 3° and the contact-angle hysteresis to 2°.

A Janus titanium mesh with superhydrophobicity and high oleophobicity was fabricated by means of electrochemical etching and wet chemistry [86]. Switchably spontaneous and unidirectional transport of both water and oil was provoked in the smart mesh by ultra-violet irradiation. Unidirectional fluid transport occurred only for fluids having specific surface tension values, and fluid selectivity depended upon the ultra-violet irradiation time. Unidirectional underwater bubble permeation through the mesh was also enabled by mesh irradiation for some 7h. A mechanism was proposed which could more reasonably explain the antigravity unidirectional transport of liquid droplets. Mesh with one side irradiated with ultra-violet light exhibited unidirectional oil-transport against gravity. The mechanism was related mainly to the asymmetrical wettability of the Janus mesh. In an idealized circular cross-section pore of a porous mesh, the Young-Laplace capillary pressure could be expressed in terms of the contact-angle between the liquid and the capillary-wall, the pore radius and the surface tension of the liquid. In the case of a wettable pore surface, liquid would be spontaneously drawn into the pore due to the positive capillary pressure. In the case of a non-wettable pore surface, liquid would be repelled from the pore owing to a negative capillary pressure. When a liquid droplet was dropped onto the unirradiated oleophobic surface, a spherical droplet would slightly penetrate the microstructure and/or nanostructure of the mesh skin. The droplet volume led to output work, and that work led to an extra pressure which pushed the liquid droplet into the pores of the contacted region. This was analogous to a liquid droplet, with increased volume and superficial area, requiring the input of extra work. A liquid droplet located on the pores became increasingly small but remained spherical due to the oleophobic surface. When its diameter was smaller than the pore size, part of the droplet would enter the pore, convex-surface forward, and be subject to pressure. The two radii of curvature of the droplet were different. An additional pressure, due to a smaller radius was helpful to transport of the droplet. When the forward curvature contacted the hydrophilic region, the flat face became concave and both of the associated pressures acted downwards, thus speeding up penetration. In the case of antigravity penetration, the outward pressure was greater than the inward pressure plus that of the liquid's own weight. When a liquid droplet was dropped onto the oleophilic surface it first spread in all possible directions due to the pull of the capillary force, and part of the liquid would enter the pore with concave surface forward. The radii of the two curvatures of the droplet are very different, due to the rapid spreading of the droplet on the superhydrophilic surface. Before the forward curvature comes close to the theoretical line across the thickness, forward and rear pressures are both close to zero. If the forward curvature makes it across the region where the contact-angle between the liquid and the capillary wall is 90°, the forward curvature becomes convex and the pressure will be upwards, thus pushing the forward curvature back into the hydrophilic region. The capillary force will pull the droplet out of the pore. When compared with previous models, this was the first to propose liquid droplet transport across the thickness from lyophobic to lyophilic gradients. The droplet is assumed to have both curvatures during penetration, and this can convincingly explain the antigravity liquid transport.

Anodic oxidation, vapour deposition and ultra-violet light were used to create a wetting-gradient and a geometrical gradient on Ti-6Al-4V surfaces[87]. The superposition of geometry-gradient and wetting-gradient was achieved by introducing a hydrophilic wedge-shaped pattern on the vapour-deposited surface. Vertically upward self-propelled motion of droplets occurred on the multi-gradient surfaces due to a synergistic interaction between the wetting-gradient force, the wetting-difference force and the Laplace-pressure force. The initial droplet velocity and distance which was moved on the vertical surface was up to 22mm/s and 14.5mm. The spontaneous motion occurred in two stages. The velocity in stage-1 was much higher than that in stage-2, due to the strong driving force and the weak pinning effect at the tip of the wedge-shaped pattern. As the inclination was increased, the distance which was moved and magnitude of the velocity decreased, due to the increasing gravitational force which acted down the inclined surface. On a horizontal multi-gradient surface, a 5μL droplet spontaneously moved by 28.7mm within 6.7s (table 10). As a droplet moved from the narrow front to the wide rear of the wedge-shaped pattern, the resistive force markedly increased, explaining the decreased velocity in the second stage.

Table 10. Spontaneous motion of droplets on various surfaces

Surface	Gradient	Driving Force	Tilt(°)	Distance(mm)	Velocity(mm/s)
A	chemical	wg	0	1-3	0.028
B	structural	wg	0	2	-
C	shape	lp	0	6	5
D	multi-	wg/wd/lp	0	2.6	-
E	multi-	wg/wd/lp	0	6.61	218.01
F	multi-	wd/lp	30	13-14	-
G	multi-	wg/wd/lp	90	14.5	22

A: graphene with fluorine gradient, B: PDMS surface with gradient wrinkle structure, C: slippery liquid-infused porous with wedge-shaped groove, D: Graphite plate with wedge shaped-pattern, E: chemical gradient aluminium surface with hydrophilic wedge-shaped pattern, F: Wettability-patterned silicon surface with nanospike-array structure, G: Ti6Al4V with wetting gradient and geometry gradient, wg: wetting gradient, wd: wetting-difference, lp: Laplace pressure

Carbon

Another bio-inspired micro-spine array imitated the energy-gradient of cactus multi-spines and the liquid-film sliding of rice leaves. Unmodified and modified array which comprised multi-walled carbon nanotubes provided ultra-fast directional transport of water and oil droplets, respectively, which was 1 to 3 orders-of-magnitude faster than that previously possible. The structural gradient of the micro-spines was the main driving force for the self-transportation, with the multi-wall carbon nanotube coating providing a uniform water film which reduced any hysteresis resistance[88]. Droplets on the array were driven by a force which was directed from the tip to the base of the array. The surfaces permitted vertical antigravity self-transport of multi-droplets, and mechanical models were developed for various droplets interacting with the micro-spine structure in order to clarify the synergy between the micro/nano-structural parameters and the chemical properties which governed droplet self-transportation. In the case of oil droplets, the forces which drove self-transportation against gravity included the Laplace force, the capillary force and the wettability-gradient force. On the other hand, water-droplet self-transport against gravity was driven by the Laplace force and the capillary force. The hysteresis resistance force between the droplet and the surface structure also added to the overall resistance to droplet self-transportation. Regardless of whether the micro-spine array surface was dry or wet, it did not exhibit superhydrophilicity and this hindered the upward transport of water droplets, due to the high resistance. On the other hand, the superhydrophilic surface of the modified micro-spine array was essential for the antigravity self-transportation of water droplets. The velocity of liquid droplets was determined mainly by the driving force acting on them. The average self-transport velocities of water droplets on modified micro-spine arrays with half-cone angles ranging from 1.4° to 2.8° were measured, and the maximum value was 237.42mm/s for an angle of 2.0°. The trend in the transportation velocities of water droplets on modified micro-spine arrays with various half-cone angles paralleled that found for oil droplets on unmodified micro-spine arrays. This was attributed to the superhydrophilic surface which had eliminated the hysteresis resistance force and increased the capillary force. The modified arrays could offer the loss-free transport of water droplets, especially those of larger volume. Water droplets with volumes of 40, 60, 80 and 100µL could be fully self-transported to the base of the modified multi-spine array within 4.76, 7.28, 9.44 and 13.48s, respectively. A very large (150µL) droplet was self-transported within 19.04s. The self-transport of multi-droplets on horizontal unmodified and modified arrays was such that average velocities of 290.91mm/s (oil) and 430.98mm/s (water) were observed. The temperature greatly affected the surface tension, thus affecting droplet self-transport. Higher temperatures also increased the self-transport velocities of oil and water droplets due to a reduced viscosity which increased flow efficiency.

Glass

The key feature of surface wettability is that it enables droplets to move uphill, and a means has been found[89] to promote the horizontal and uphill self-motion of tetrahydrofuran droplets in the absence of an external force. Non-aqueous tetrahydrofuran drops which contained hexamethyldisilane were placed on a completely wetted glass surface. They underwent uphill motion at an angle of some 10°. The binary drops were also able to convey additional co-solutes. When droplets exhibit antigravity motion, their motion is generally driven by directional wetting. This occurs when the contact-angle of a droplet on a tilted surface is asymmetrical and causes the droplet to move in a specific direction contrary to the gravitational force. The direction of a moving droplet is governed by the wetting properties of the droplet and the surface: if the contact-angle is greater on one side of a droplet than on the other, the contact-line tends to move recede from the side having the smaller contact-angle. When there is an asymmetry in the surface energy or surface chemistry, the droplet can experience a nett force which propels the droplet against gravity, but contact-angle hysteresis can cause droplet-pinning. When the hysteresis is small, a droplet tends to exhibit directional motion that is caused by surface roughness, a wetting gradient, a chemical gradient or anything which leads to an asymmetrical wetting behaviour. On a tilted high temperature treated glass surface, the uphill motion was most marked for a 1:1 concentration of hexamethyldisilane and tetrahydrofuran. The volumetric ratios of hexamethyldisilane to tetrahydrofuran were 0.5:1, 0.75:1, 1:1 and 1.25:1. All the concentrations exhibited drop motion, and the velocities were 2.68, 4.17, 4.65 and 5.83mm/s, respectively. The antigravity motion was attributed to interfacial-tension imbalance. The wetting-area of droplets exhibited fluctuations during motion; gradually decreasing due to evaporation of the volatile component and involving dynamic spreading and receding behaviours. The driving force for uphill motion was linked to altered interfacial tensions, as affected by changes in the surface energy and partial chemical composition. The presence of 6mM of dye caused contact-line pinning and altered droplet motion. Lower concentrations permitted drop motion on treated glass surfaces.

Graphene

A simple approach was proposed[90] for the creation of an open micro-fluidic device which manipulated the wettability of spin-coated graphene-ink films on flexible polyethylene terephthalate via laser-controlled patterning. Wedge-shaped hydrophilic tracks which were surrounded by superhydrophobic walls were produced in graphene films by scribing micron-sized grooves onto the graphene via CO_2 laser. The scribing process produced superhydrophobic walls having a water contact-angle of about 160°. These walls enclosed hydrophilic tracks which were created by oxygen-plasma pre-treatment of the graphene. These all-graphene microfluidic tracks were able to transport droplets at a velocity of 20mm/s on a level surface. They could also transport droplets uphill at an angle of 7°. Such devices could be created by using exfoliated flakes of graphene without advanced processing. The patterned graphene retained dynamic superhydrophobic properties for 15 days.

A laser-induced graphene/polyimide slippery membrane was examined[91] which exhibited a good photothermal response and good oil absorption. The application of local near-infrared irradiation could drive both droplet and bubble movement on the membrane. As well as horizontal manipulation of droplets and bubbles on the slippery laser-induced graphene/polyimide membrane, antigravity droplet movement and anti-buoyancy bubble transport could be driven by the irradiation. A 5μL water droplet could be caused to move uphill against gravity under irradiation of a membrane with an inclination of 3° (and 1 centistoke silicone oil). The average sliding velocity of the droplet was about 0.375mm/s over 20s. The antigravity movement also led to the coalescence of droplets on the inclined membrane under irradiation. A 2μL water droplet first moved downwards, under gravitation force, and a 5μL water droplet moved upward under irradiation on an inclined (3°) membrane. These droplets coalesced to become a larger droplet after 4s and continued to move uphill irradiation at an average sliding velocity of about 0.35mm/s. The critical antigravity sliding angle of a 5μL water droplet on the membrane was about 8°. A bubble in an underwater environment could move downhill, against buoyancy, on an inclined membrane (1 centistoke silicone oil), and the critical angle for a 5μL air bubble to move downhill was about 3°. The forces which acted on the droplet were fourfold: these were the gravitational force, the positive supporting force between the droplet and the membrane, the wettability-gradient force and the resistive force. In order that the droplet move uphill, it was necessary that the wettability-gradient force minus the resistive force minus the gravitational force, multiplied by the sine of inclination angle, be greater than zero. In the case of a bubble moving downward on the inclined membrane, the relevant forces acting on the bubble became the buoyancy force acting on the bubble, the adhesion force between the bubble and the membrane, the wettability-gradient force and the resistance force. These forces then had to satisfy the condition that the wettability-gradient force minus the resistance force minus the buoyancy force, multiplied by the sine of the inclination angle be greater than zero. When these conditions were satisfied, the critical inclination angle for a droplet climbing upward or for a bubble moving downward could be calculated. The thus-predicted angle for a 5μL water droplet was about 15° and the angle for a 5μL air bubble was about 7°: both agreed with experimental data. The main reason for antigravity droplet movement and anti-buoyancy bubble transport was deemed to be the dominant wettability-gradient force which resulted from the good photothermal properties of the membrane.

Silicon

The spontaneous motion of droplets was studied[92] (figures 5 and 6) on patterned silicon substrates having etched micro-channels with a height of 0.4μm and a width of 10μm and a space of 10μm between them. A gradient in surface energy. The water contact-angle of the photo-degraded siloxane could be adjusted. Surfaces was produced by photo-degrading a thin layer of octamethylcyclotetrasiloxane which had been deposited onto the patterned substrate by means of plasma-enhanced chemical vapour deposition with contact-angles ranging from 104° to 22° for water droplets were obtained. Deionized water droplets with a volume of 60μL were used. The combination of surface-energy gradient and etching improved droplet displacement over that of smooth surfaces. The effect of gravity was studied by tilting surfaces by up to 20°. When horizontal, the peak velocity attained some 60mm/s on the patterned substrates with the direction parallel to the micro-channels.

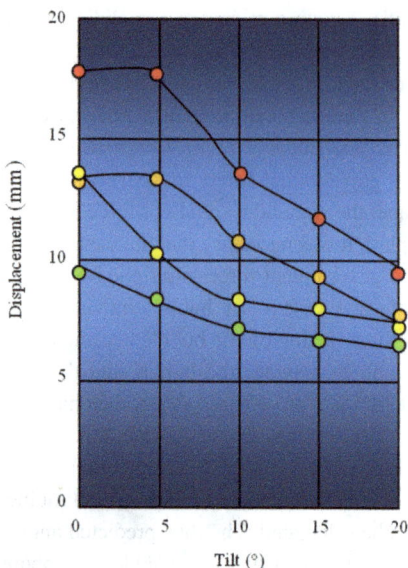

Figure 5. Effect of substrate inclination on the displacement of droplets. Red: advancing edge (grooved surface), orange: centre-of-mass (grooved surface), yellow: advancing edge (smooth surface), green: centre-of-mass (smooth surface)

It reached some 64 on smooth substrates. The horizontal displacement attained 18mm on the patterned substrate and 14mm on the smooth substrate. In the case of 20° inclined substrates, the peak velocity decreased to 10.2mm/s for patterned substrates and 22mm/s for smooth substrates. The displacements meanwhile decreased to 9.6mm for patterned substrates and 7.6mm for smooth substrates. Droplets were able to move spontaneously uphill on vertical surfaces with micro-channels having a height of between 1 and 2μm. The peak velocity attained 7mm/s for a total displacement of about 8mm. The micro-channels could have a positive impact upon the displacement of the droplets on both horizontal and inclined substrates. The micro-channels constrained the droplets to remain almost constant in width in the direction perpendicular to their motion. By adjusting the gradient and the depth of the micro-channels it was possible to make droplets spread uphill on a vertical grooved surface (figures 5 and 6). The depth of the micro-channels was increased in order to prevent the droplet from spreading laterally. When the centre-of-gravity of a droplet moved uphill, the front edge of the droplet moved contrarily to the back edge, leading to a directional spreading. Increasing the depth of the micro-channels permitted uphill motion of the front edge, but increased the effect of a hysteresis of the contact-angle.

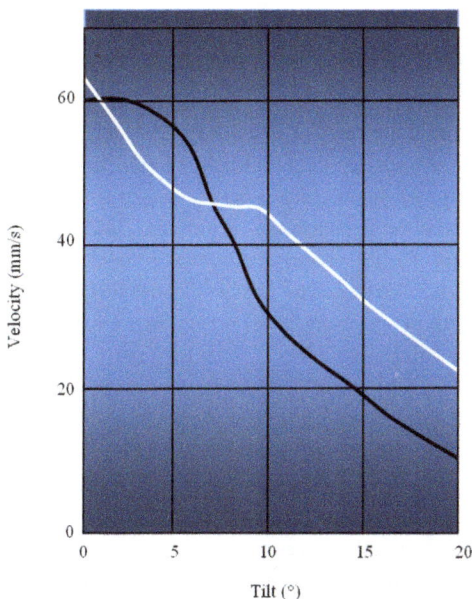

Figure 6. Effect of substrate inclination on the velocity of droplets. Black: centre-of-mass (patterned surface), white: centre-of-mass (smooth surface)

Experimental studies and theoretical analyses were made[93] of the adhesive forces between captive air-bubbles (8 to 20μL) and superhydrophilic surfaces, and were compared with those for hydrophilic surfaces. Polished bare silicon substrates and micro-pillar silicon substrates, with contact-angles for sessile water droplets of 40° and 4°, were used as the hydrophilic and superhydrophilic surfaces, respectively. The adhesive force of an air-bubble to an inclined surface was predicted to depend upon the contact-width and the contact-angle hysteresis. The latter was defined, as usual, to be the difference in the contact-angles between uphill (advancing) and downhill (receding) sides. It was shown that the sliding-angle, contact-angle hysteresis and adhesion force were much lower on the micro-pillar superhydrophilic surface than on the hydrophilic surface. This was due to the presence of an entrapped water-layer on the micro-pillar superhydrophilic surface. For both surfaces, the adhesive force decreased slightly with the bubble-volume. The contact-width increased with increasing bubble-volume but the contact-angle hysteresis decreased more markedly. The adhesive force therefore effectively decreased with increasing bubble-volume. This was attributed to the buoyancy force, which increased with increasing bubble-volume and depended upon the shapes of the air-bubbles which formed on surfaces with differing wettabilities.

An additive manufacturing method has been proposed[94] for the creation of silicon-based heterogeneous functional surfaces that could manipulate droplets. It permitted the instantaneous formation of hierarchical multi-scale structures which offered tunable wettability. This involved the instantaneous plasmonic sintering of silicon particles with no need for masking or further processing. It was possible to combine heterogeneous surfaces into a single domain by reversibly switching the wettability from superhydrophobic (with a contact-angle of 161.2°) to superhydrophilic (with a contact-angle close to 0°) via laser irradiation. The existence of continuous superhydrophilic channels, with a contact-angle of 9.6°, created on the superhydrophobic silicon background by re-irradiating with a laser beam, led to droplet motion along the channel under the action of van der Waals forces and Laplace pressure fields which were generated by the difference in wettability. A water droplet which fell onto the surface slid along the superhydrophilic channel, under confinement by van der Waals forces. A droplet which was wetted on wedge-like superhydrophilic channels, with differing Laplace pressures at the front and rear interfaces, could propel itself. If the nett Laplace pressure force was higher than the vector sum of gravity and surface friction, the droplet could climb an inclined channel without the aid of an external force. A droplet resisted sliding perpendicularly to the channel direction. The difference in sliding angle in the channel direction increased sharply as the channel interval decreased. That is, an angle-difference of 5° over a 0.2mm interval, increased to an angle difference of 32.7° over a 0.1mm interval. This showed that the surface tension could impede or assist the motion of a water droplet, depending upon the angle between the gravitational force and the superhydrophilic channel. The sliding motion of water droplets could thus be controlled by adjusting the tilt angle of the substrate, or the relative angle between the channel and the gravitational force. A deeper study of droplet control showed that the meniscus curvature on each side of a droplet, along the channel direction, could be made different by continuously adjusting the width of the superhydrophilic channel. The droplet could then be self-propelled by the spatially-modulated difference in Laplace pressure with no assistance from gravity; it could even move against gravity. The Laplace pressure caused

Materials Research Forum LLC

https://doi.org/10.21741/9781644903896

by the surface tension of the interface between air and water was the difference between the inside and the outside of the meniscus surface.

Two novel methods were described[95] for reliable contact-angle measurement on inclined high-quality hydrophilic silicon wafers which were synthesized by silanization with 3-aminopropyltriethoxysilane. In these so-called slow-moving contact-angle analyses, the average curved shape of the contact-angle relative to the inclination angle was represented by a sigmoid function, or a specific contact-angle analysis was used to determine an advancing or receding angle, leading to specific distributions. Scatter in the data was attributed to a combination of preparation vagaries and the effects of the bulk material on the contact-angle. Differences in the formation kinetics of the siloxane layer in the presence or absence of additional water and acid suggested that a well-ordered defect-poor and a less-ordered thin layer of mono-(3-aminopropyl)siloxane were produced. Atomic force microscopy revealed only slight and insignificant differences in the standard deviations for the two surfaces. Unrecognised differences could lead to unexplained effects in adhesion. Using inclination/contact-angle analysis, the surface could be investigated with high sensitivity and fine local resolution. The high quality of the surfaces and the ability to obtain valid values via global analysis was confirmed by small differences in the resultant values for acceleration/deceleration events, with 56.7° (downhill) and 40.73° (uphill) for the contact-angle and with 13.03° (downhill) and 14.9° (uphill) for the inclination angle. The analysis resulted in characteristic density distributions of the contact-angles and showed that the motion of the triple-line on an inclined surface was very complex. The distributions for the uphill angles were not independent of the downhill angle. Due to this, the downhill and uphill angle which were obtained by inclining the surface did not correspond to the advancing and receding angles which were measured on a horizontal surface. Because the apparent contact-angles were affected by the local properties of the surface, a correlation between the surface properties and the density distributions existed. Determining just one advancing and one receding angle via static measurements, in order to characterize surface properties, was critical. The usual procedure for measuring the uphill and the downhill angle on the same image was deemed to be incorrect. The values which were determined for these angles were greatly affected by the type of optical observation and by the temporal and spatial resolutions of the measurements. It was concluded that sessile-drop experiments were superior for obtaining reproducible and consistent contact-angle measurements. These techniques could even identify inhomogeneities in the surface tension of very thin and very flat layers which were required in order to explain adhesion effects.

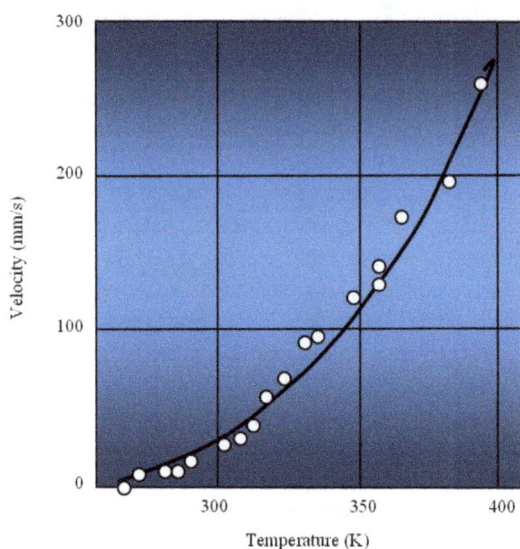

Figure 7. Relationship between the acid droplet velocity on silicon surfaces and the ambient temperature

Hydrofluoric acid etching of a silicon surface is an efficient method for creating interfacial chemical reaction flow and the consequent self-propulsion of water droplets. Etching causes a large surface free-energy change, as reflected by an increase in the water contact-angle from 0 to 60°. This favours self-propelled hydrofluoric acid droplet motion at high speeds over long distances. In particular, it permits vertical motion of a 10μL acid droplet on a silicon strip. Investigation[96] of the temperature-dependence of the velocity shows that the velocity is proportional to the acid reaction-rate. When a surface having a chemical gradient was first constructed[97] using chemical vapour deposition, 1 to 2μL water droplets could exhibit uphill motion on a modified surface which was inclined at 15°. A droplet of decane which contained a fluorinated fatty acid[98] was able to run up an incline of 43°. The relationship between reaction-rate and droplet velocity was determined by investigating the effect of the ambient temperature on the velocity (figure 7). The uphill motion of 2μL acid droplets on various inclines was measured (figure 8). The droplets easily propelled themselves even when the silicon surface was vertical, and the driving force arising from the interfacial chemical reaction was therefore large enough to overcome gravity. The effect of gravity did not dominate in the case of smaller droplets. Upon increasing the droplet volume, there was a maximum velocity for the various angles of inclination. The critical droplet-size which corresponded to zero velocity could not be obtained experimentally, but extrapolation of the plotted data suggested that 10.6µL was the critical droplet size for vertical motion. This suggested that a 3mm-wide silicon strip could theoretically support the vertical climb of a droplet having a maximum volume of 10.6µL. Vertical uphill motion of a 10µL acid droplet on a 5mm-wide silicon strip was demonstrated, with the droplet travelling from the bottom to the top.

Zhang W., Dai X., He S., Guo Y., Guo Z., Chemical Engineering Journal, 482, 2024, 148928.

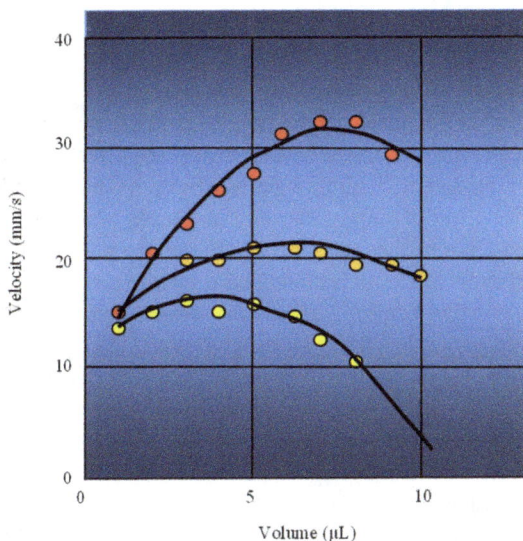

Figure 8. Relationship between acid-droplet velocity and droplet volume on inclined silicon surfaces. Red: 15°, orange: 45°, yellow: 90°

The determination of the contact-angle of sessile droplets shows that various angles can be distinguished which are correlated with advancing or receding of the triple line, and

Materials Research Forum LLC

https://doi.org/10.21741/9781644903896

these may not be clear or reproducible. The motion of droplets on flat silicon-oxide surfaces, following divers surface treatments, was measured[99] dynamically on inclined specimens. Triple-points, inclination angles, downhill (advancing motion) and uphill angles (receding motion) were analysed. The methods were applicable to static or slowly-moving drops. The distribution of values for the uphill angle was not independent of the downhill angle. There was a bimodal distribution of the downhill contact-angle as a function of the inclination angle. The contact-angles were essentially non-moving between -4 and 22μm/s and were unaffected or little affected by dynamic effects. Three ranges could be defined: non-moving (static), slow-moving and dynamic (high velocity, sliding). All of the measurement methods were expected to be less influenced, by hysteresis of the contact-angle than were purely static contact-angle determinations. The problem of contact-angle hysteresis (advancing and receding angles), as studied on inclined planes, was concluded to remain unsolved and it was suggested that only statistical analysis could lead to reliable contact-angle determinations.

Inclined nanoforests which possessed superhydrophilic properties were created[100] via a mask-free route which required just subsequent oxygen and argon-plasma treatment. Droplets on the surface underwent asymmetrical spreading at room temperature. Upon heating the surface to below the Leidenfrost point, and letting-fall a droplet onto the surface, directional motion of the droplet occurred within a transition-boiling state. This movement was due to an asymmetrical surface tension which was produced by the inclined nanoforests. The surface temperature affected the rate of horizontal propulsion, which occurred mainly in the transition-boiling state where the instantaneous phase transition and intermittent contact between the surface and the droplets led to large variations in droplet propulsion. The average horizontal velocity depended upon surface temperature and tilt-angle. At a temperature of 260C, the average horizontal droplet velocity gradually increased for Weber numbers ranging from 1.7 to 4.0, and peaked at 4.0. As the Weber number increased further, evaporation of the droplets led to a reduced horizontal momentum. When the temperature was 380C, the average horizontal droplet velocity reached a peak at a Weber number of 4.6. In addition to moving horizontally, the droplets could also move uphill. As the surface temperature increased from 260 to 280C, the uphill velocity of droplets decreased. Evaporation markedly affected the stability of uphill motion. At 260 to 280C, droplets with smaller Weber numbers easily underwent directional propulsion.

Acoustofluidic methods are generally restricted to the manipulation of droplets on flat surfaces. An attempt was made to extend their use to inclined surfaces for the purpose of microfluidic device design. Theoretical and experimental studies were made[101,102] of acoustofluidic transportation/pumping and jetting on inclined surfaces using AlN/Si

Rayleigh surface acoustic waves. Droplets having a volume of less than 3μL could be transported on various inclined surfaces, with gravity obviously playing an ever-increasing role in uphill climbing with increasing inclination angle. When the inclination was 90°, a higher threshold power was required in order to transport a droplet, and the maximum droplet-volume which could be pumped also approached its minimum value. The effect of inclination-angle upon droplet jetting-angle could be neglected for volumes of less than 2μL. When a radio-frequency power of 54W was applied, the acoustic streaming forces from two sides pushed the droplet upwards to form a vertical cylindrical liquid beam and liquid was then ejected from the device surface. For a 1μL droplet, the jetting-angle for various inclinations were almost vertical to the device surface. The jetting-angles of 2μL droplets on the flat and inclined surfaces were also perpendicular to the device surface, but the liquid beam on an inclined surface tended to bend downwards at the tip of the elongated beam, due to gravity. In the case of AlN/Si surface acoustic waves, acoustic streaming occurred at a radio-frequency power of just a few mW. The threshold-power which was required to pump the droplet was a few watts. As the inclination angle increased from 0 to 90°, the threshold pumping power increased to a maximum value of 10.5W. A higher power resulted in the ejection of a droplet from the device surface and the minimum threshold jetting-power of 30W for a 1μL droplet occurred at an angle of 180°.

Polymers

A droplet-based microfluidic system was developed[103], for biochemical assay, which permitted the manipulation of droplets via self-mixing and spontaneous movement. The latter driving force arose from a gradient in surface tension which was generated by a hydrophobic micro-textured polydimethylsiloxane surface. This included uphill migration. The use of a surface-energy gradient to generate droplet movement is commonly exploited via electrowetting and chemical deposition. Surface roughness was here produced on a micro-textured polymer chip, and this roughness governed the hydrophobicity which was generated by a nett pressure-difference between the ambient air and a super-hydrophobic surface. The nature of the contact-interface between the droplet and the ambient air was governed by the equilibrium of surface energy, according to the Laplace-Young equation,

$$\Delta P = \gamma(1/r_1 + 1/r_2)$$

where ΔP was the difference between the internal pressure and the external pressure of the spherical surface of the droplet, γ was the surface tension of the droplet and the r were the two radii-of-curvature of the surface. A droplet moved from a super-hydrophobic surface to a hydrophobic surface. By creating artificial surface roughness on a polymer micro-textured biochip it was possible to facilitate analysis.

There exists a temperature-dependent tunability of the wettability of poly(N-isopropylacrylamide) surfaces below and above a lower critical solution temperature of 32C. The transport of water droplets is however limited by a large contact-angle hysteresis. Rapid reconfigurable droplet manipulation over a poly(N-isopropylacrylamide) grafted structured polymer surface was possible[104] by introducing a temperature-induced wettability gradient. Poly(N-isopropylacrylamide) which was grafted onto intrinsically superhydrophobic surfaces had a hydrophilic nature, with a contact-angle hysteresis below 30C, and a superhydrophobic nature with ultra-low contact-angle hysteresis above 36C. The transition between 30C and 36C was associated with a large (circa 100°) change in water contact-angle, together with a related change in contact-angle hysteresis. This created a so-called transport-zone within which driving forces overcame frictional forces. The macroscopic transport of water droplets at a maximum velocity of some 40cm/s was demonstrated. Theoretical predictions of the force agreed with the behaviour of driving forces across the transport-zone. The transport velocity could be tuned by varying the temperature gradient along the surface, or the inclination angle of the surface (table 11). A maximum angle of 15° led to a velocity reduction of 0.4mm/s. The coalescence of water droplets was also possible by exploiting the temperature-controlled wettability gradient. In some tests, the smart textured surface was kept at various inclinations, with the temperature gradient in the uphill direction and the hot side down. The steepness of the temperature gradient was kept the same as that for horizontal surfaces. The droplet climbed, but with a reduced (gravity-limited) velocity (0.37mm/s) when compared to that in the horizontal situation. As a control process, experiments were carried out at fixed temperatures which were below and above the lower critical solution temperature while keeping the inclination angle at 15°. In the former case, the droplet spread over the surface. In the latter case, it rolled off in the direction of the gravitational force; the direction opposite to that of wettability-driven transport. The results confirmed that droplet-motion in the presence of a temperature gradient is due to the wettability-gradient forces. At inclinations of 20° or more, motion occurred in the direction of gravity, indicating that the wettability-gradient was insufficient to move the droplet upward. With the temperature gradient and the inclination kept the same as in the case of uphill transport, two droplets were positioned, one on the cold side and the other on the hot side. The droplet which was on the cold side spread because of the low water contact-angle at that temperature. The droplet placed on the hot side experienced the wettability-gradient and moved towards the other droplet,

Materials Research Forum LLC
https://doi.org/10.21741/9781644903896

and merged. In order to distinguish this from gravity-driven motion, the experiments were performed at an angle of about 2°. Droplet-motion clearly occurred uphill.

Table 11. Droplet velocity at various temperature gradients

$T_{hot}(C)$	$T_{cold}(C)$	Gradient(C/mm)	Velocity$_{maximum}$(cm/s)	Velocity$_{average}$(cm/s)
36	32	1	43.1	9.15
35	32	0.75	33.05	11.34
34	32	0.5	23.4	14.4
33	32	0.25	11.9	8.01

A layer of lubricant on an underlying rough matrix paradoxically enables behaviours which are impossible on superhydrophobic surfaces. A type of slippery material was developed which could exploit the advantages of both solid and lubricant. This was possible by constructing[105] a photothermal-responsive composite matrix which permitted real-time light-induced surface-charge regeneration and allowed the photo-control of droplets. This light-induced charged slippery surface permitted the photo-control of droplets undergoing rapid long-distance antigravity and directional motion. Unlike conventional slippery lubricant-infused porous surfaces, the light-induced charged slippery surface consisted of three components, including (1) micro-size Ga-In liquid-metal particles which efficiently converted absorbed light into localised heat, (2) poly(vinylidene fluoride-co-trifluoroethylene) copolymer which exhibited ferroelectric behavior and (3) microstructures which were coated with a layer of hydrophobized silica nanoparticles for trapping lubricant. As compared with passive slippery lubricant-infused porous surfaces, the light-induced charged slippery surface exhibited a better light-induced droplet manipulation which included a high velocity (~18.5mm/s), long distances (~100mm) and antigravity motion for droplet volumes of 10^{-3} to 1.5 x 10^{3}μL.

Droplet transportation in shape-memory polymer tubes was achieved mainly by exploiting the tube's elastic shape-variation, with a pitcher-plant like slippery surface coating being required to modify the inner surface and ensure complete wetting by various carried liquids. The slippery surface reduced adhesion and extended the possible range of liquids transported, with water and oils which possessed a wide range of surface tensions being controllable. The transportation speed and direction could be modified[106], and included antigravity motion. The shape-memory polymer provided a variable tubular shape and the slippery surface reduced adhesion.

Corona discharge and contact electrification were combined[107] in order to move droplets on a superhydrophobic surface. The marked adhesive effect of contact electrification caused droplets to stick to a slope without sliding down when the discharge ended. The discharge could easily propel the droplets, and also quickly charge the polymer surface. Uphill, zero and downhill motion could be obtained by combining the electrostatic force due to the charge on the polymer surface. The droplet's gravitational potential energy could be stored and released. A control system was constructed in which the actuator consisted of an aluminium plate-electrode and a stainless-steel needle-electrode. The aluminium electrode was grounded and the stainless-steel electrode was connected to a negative supply. The vertical distance between the needle and the substrate was about 5mm. A 2nm superhydrophobic layer of polydimethylsiloxane was placed on the aluminium electrode. The static contact-angle of a 5μL was about 155°. A droplet having a volume of about 10μL was initially placed on the layer. When the voltage which was applied to the needle was greater than 2kV, a coronal discharge occurred near to the tip and the air was ionized, with the positive ions being absorbed by the negatively charged needle while the negative ions were absorbed by the droplet and the droplet surface acquired some negative charge. The use of a higher voltage did not mean that the droplet immediately acquired a higher velocity. A higher voltage increased the frictional resistance and droplet propulsion became more difficult. When the applied voltage was close to 8kV, the increased frictional resistance was such that the droplet could not be accelerated in the early stages. When the droplet moved away from the needle-electrode, the frictional resistance decreased and the droplet accelerated to velocities as high as 30cm/s. If the droplet was far from the needle, a higher voltage was required in order to move it. For a given applied voltage, there was therefore a maximum distance over which the droplet could be moved. In experiments, that distance was affected by the droplet-size and the applied voltage. The greater the droplet radius, the more the electric charge acquired so that the maximum distance was greater at that same voltage.

Natural surfaces, such as water-repellent lotus leaves and superhydrophobic water-adhesive rose petals suggest that hierarchical structures are essential for controlling water behaviour. Superhydrophobic and anti-reflection silicon nanospike-array structures were created[108] by using self-organized honeycomb-patterned films as 3-dimensional dry-etching masks. The surface wettability of silicon structures could be easily changed from superhydrophobic to superhydrophilic by making changes in the surface chemistry. Honeycomb-patterned films for dry-etching masks were made from polystyrene and an amphiphilic polymer by casting a chloroform solution. Reactive ion etching was

Materials Research Forum LLC

https://doi.org/10.21741/9781644903896

performed following fixation of the top layer of the honeycomb-patterned film on a monocrystalline silicon substrate. The as-prepared silicon nanospike-array structures exhibited superhydrophobicity, and the water contact-angles were greater than 170°. Following UV-O₃ treatment with photomasks, only ultra-violet exposed surfaces exhibited superhydrophilicity. This suggested that superhydrophobic and superhydrophilic patterned surfaces could be created in which the patterns were the same as those of the photomasks. Such wettability-patterned surfaces were used to demonstrate water-harvesting by superhydrophilic dot-patterned surfaces, and uphill water transportation by superhydrophilic triangle-patterned surfaces. A triangular shape was essential for water transport because the interfacial energy increased with increasing superhydrophilic area. The angle of the triangle was 10°. Only the UV-O₃ irradiated triangular area was wetted. The substrate was tilted and a hydrophilic glass slide was placed upon it. When droplets were dropped onto the inclined wettability-patterned surface, the first droplet adhered to the surface and climbed slowly as the proportion of superhydrophilic surface increased. The droplet then remained on the upper part, where the interfacial force balanced the effect of gravity. When a second droplet was dropped onto the surface, it immediately merged with the first droplet and the position of the centre-of-mass moved upwards. As the water droplet increased in size, it eventually reached the hydrophilic glass slide, where it then spread over the surface of the glass slide. From the third droplet onwards, the water rapidly climbed the inclined surface and spread over the surface of the glass slide. In order to investigate further the uphill water transport, measurements were made of changes in the water contact-angle and the wetting-distance of the leading edge of the water. The inclinations were 0°, 15° and 30° and the droplet volumes were 2.0, 5.0 and 8.0μL, respectively. On the basis of the experimental data, an equation of motion was derived for the uphill motion of water. The overall theory of uphill motion of water droplets on superhydrophilic inclines first considered the equation of motion of the droplet, which included the driving force at the leading edge of the climbing droplet and the opposing forces of gravity, pinning and energy dissipation due to viscosity. The pinning effect acted on the tail of the climbing droplet at the boundary between the superhydrophilic and superhydrophobic areas.

Superhydrophobic micro-tower arrays were created[109] in order to study spontaneous uphill movement, a Wenzel-to-Cassie transition and the self-removal of condensates. The arrays permitted the spontaneous uphill movement of tiny condensates, which were trapped in microstructures, due to an upward Laplace pressure which was some 30 times greater than that on cone-like arrays. The presence of sharp tips decreased adhesion to the suspending droplets and hastened their rapid self-removal. Such condensates formed in rough structures due to the nucleation of vapour on an homogeneous cold surface. When the entrapped droplets failed to move up from the valley to the peak of a microstructure or nanostructures, they were in the Wenzel state. This increased the surface adhesion to condensed droplets. The promotion of spontaneous uphill motion movement of a condensate which was trapped in the valley of a microstructure or nanostructure was essential to a wettability transition and consequent self-removal. The present tower-like-array surfaces allowed the rapid self-removal of condensates because the sharp tip of an array decreased the solid/liquid contact area and the surface adhesion to condensed droplets. In order to investigate the effect of the side-wall morphology upon uphill movement, polydimethylsiloxane was used to obtain various side-wall morphologies. The sharp tip of a tower-like array led to a discontinuous unstable 3-phase contact-line which further decreased surface adhesion and promoted the rapid self-removal of condensates.

A magnetically responsive superhydrophobic surface was prepared[110] by means of 2-step spraying. The surface provided the accurate and rapid control of droplets by magnetic means. Single droplets could be moved along unidirectional, circular, irregular and inclined planes using no additional energy. The surface could grasp and release droplets by controlling the magnetic field. The effects of various Fe_3O_4 mass ratios on the sprayed surface's morphology and wettability were determined and a model of droplet control was used to explain the mechanism of droplet behaviour. The materials used included polydimethylsiloxane and Fe_3O_4 with an average particle size of 30μm. The magnetic field was provided by a neodymium permanent magnet having a field strength of 1.5T. In order to check that the superhydrophobic surface exhibited low adhesion, a droplet was released and rolled freely. A 30μL droplet was freely released by an inclined surface with an angle of 15°, and rolled off quickly with no adhesion to the surface. There was thus low adhesion in the absence of a magnetic field. When the magnet was placed below the substrate, a droplet that fell in that area was controlled because the magnetic field-lines were non-uniformly distributed and the magnetic force in the central region was weak. There would be a potential difference between a droplet in the central area and the surrounding area. The potential difference disappeared when the magnet was removed and the droplet rolled downward. The droplet retained its spherical shape. The transport of a droplet against gravity on an inclined surface was possible by using the magnet. In the case of a surface tilted at about 30°, the droplet moved from the bottom to the top upon moving the magnet. The droplet resistance increased on the inclined plane but the magnet overcame gravity.

A method was described[111] for creating a hydrophobicity gradient on the surface of a polydimethylsiloxane dry adhesive. It required partial silanization of the surface of the adhesive by means of the chemical vapour deposition of octadecyltrichlorosilane. The partial silanization of the surface produced a hydrophobic to hydrophilic gradient across the surface. The resultant change in hydrophobicity across the surface then resulted in uphill motion of a water droplet, and this appeared to be directly proportional to the contact-area between the droplet and the adhesive. Although there was a variation in the adhesion strength across the sample, the adhesive properties were barely affected by the silanization and the motion of water droplets did not cause any loss of adhesion. The same treatment was applied to flat and structured substrates and, in both cases, uphill motion of water droplets occurred. There was no difference in the velocity of water droplets across the surface of flat and structured dry adhesives. The size of the droplet did however affect droplet motion because a larger contact-surface area experienced a higher hydrophobicity gradient. A minimum droplet-size of 4µL was required before movement could occur.

A biomimetic lubricant-impregnated slippery surface was explored[112] for the manipulation of droplet transportation by using a light-based method involving a photosensitive lubricant. Localized heating using a focused laser beam, via photothermal conversion, induced Marangoni flow. The photosensitive lubricant-impregnated slippery surface provided very stable droplet transportation under the control of the laser beam. This was attributed to the stable interface temperature gradient which governed transportation. As well as droplet transportation on the flat surfaces, a 5µL water droplet could be moved uphill on a surface which was inclined at 5°. The droplet initially moved downward on the incline under the action of gravity. Upon laser irradiation, the droplet velocity gradually decreased due to Marangoni flow and an associated localized photothermally-induced temperature gradient. The droplet movement finally reversed and it moved uphill. The direction of droplet motion could be closely controlled by varying the location of the laser irradiation, and could also provoke droplet coalescence. When the laser beam was directed near to the droplet edge, the latter began to move in the opposite direction to that of the laser beam. Under continued irradiation, the moving droplet eventually came into contact with a target droplet. After a certain time, the two droplets overcame the barrier of a silicone-oil layer and coalesced. Directional remote control of a two-component (1.5wt%H_2O_2) droplet was also possible, and this droplet could be moved to a target location at which MnO_2 particles were dispersed in the infused silicone-oil layer. When the peroxide-containing droplet contacted the MnO_2 particles, the peroxide rapidly decomposed and produced large numbers of oxygen bubbles. The lubricant-impregnated slippery surface was created by spin-coating a thin layer of silicone-oil onto a polydimethylsiloxane substrate containing embedded Ti_2O_3 particles. The detailed mechanism was therefore that, under laser irradiation to one side of a droplet, a temperature gradient was created by the localized heating, via photothermal conversion, leading to thermal Marangoni flow within the infused oil layer and the droplet. The oil layer drained from the laser irradiated point, thus forming a concavity. This resulted in the creation of asymmetrical wetting ridges on each side of the droplet. The oil could then creep upward, along the air/droplet interface, under the influence of Marangoni flow within the droplet and finally enwrap the droplet so as to form an emulsion. The thermal Marangoni flow and the asymmetrical horizontal component of the surface tension, resulting from the asymmetrical ridges thereby actuated droplet transportation. The thickness of the oil layer dominated the asymmetrical horizontal component of the surface tension, and directly governed the driving force and droplet velocity.

A silicone-oil infused magnetic porous polydimethylsiloxane surface which allowed magnetically tunable super-wetting was described[113]. The surface permitted a magnetic field to control the seepage and absorption of silicone oil in micropore channels so as to modify the adhesion of water droplets to the surface. The latter underwent a reversible *in situ* conversion from ultra-high to ultra-low droplet adhesion, and the sliding angles ranged from 3° to 180°. Droplet manipulation on inclined surfaces generally involved using gravity as the driving force. This was not possible in the case of horizontal or vertical surfaces. When an incline was less than 5°, droplet manipulation was difficult in the absence of external driving force, due to a low gravitational driving force or a high surface adhesion. Upon introducing photothermal, magnetic or electric forces, droplet manipulation on a horizontal surface became possible, as well as transport on a slightly-inclined surface. Precise manipulation required ultra-low adhesion, but stable manipulation required ultra-high adhesion. The silicone-oil infused magnetic porous polydimethylsiloxane surface conveniently offered both of these properties and thus a great advantage in micro-droplet manipulation. A water droplet was deposited on a platform which consisted of two facing silicone-oil infused magnetic porous polydimethylsiloxane samples and a magnetic field was applied directly below the platform so as to constitute an ultra-low adhesion surface. The upper platform was mobile and offered ultra-high adhesion. Due to the difference in adhesion, one force was greater than the force exerted, by the platform on the droplet, in the horizontal direction. When the upper platform was moved, the droplet was pinned to the upper platform and moved. When the magnetic force was applied directly above the platform, the droplet was pinned to the lower platform. The water-adhesion difference between the upper and lower platforms, in the horizontal direction, thus provided a precise means for directional droplet transport. This silicone-oil infused magnetic porous polydimethylsiloxane platform could thus transport various droplets to specific positions. The system could also be used for antigravity droplet manipulation on inclined, and even vertical, surfaces. When the platform was inclined, the adhesion force for droplet motion had to be equal to,

Materials Research Forum LLC

https://doi.org/10.21741/9781644903896

or greater, than the sum of the wall resistance and the gravitational component of the droplet in the direction of motion. The maximum droplet volume which could undergo antigravity motion was about 7µL when the platform was vertical. In a particular case, 3µL droplets of KSCN aqueous solution were dragged upwards so as to react with $Fe_2(SO_4)_3$ droplets.

The horizontal and uphill movement of water droplets was driven simply by patterning shape-gradient hydrophilic materials such as mica on a hydrophobic matrix such as wax or polyethylene. Such a shape-gradient composite surface was best for propelling droplets at a high velocity and for greatest distances on various geometrical composite surfaces. The driving force arose from the large difference in the wettabilities of the surface materials, from the low contact-angle hysteresis of the surface materials and from the space limitation of the shape-gradient transportation area. The average velocity and the maximum travel distance were governed mainly determined by the gradient-angle, the droplet-volume and contact-angle hysteresis. A high contact-angle hysteresis could decrease the average velocity[114]. When the gradient angle was increased from 0° to 45°, the average velocity increased from 2.4 to 15.2cm/s, while the maximum distance travelled decreased from 26 to 4mm. When the droplet volume was increased, the average velocity always decreased. On horizontal mica/wax and mica/polyethylene composite surfaces, the water droplets could run their full course even when they were as large as 50µL. With increasing volume, the average velocities decreased from 10.0 and 6.8cm/s to 1.8 and 0.6cm/s, respectively. On composite surfaces which were inclined at 25° to the horizontal, the droplets could not run uphill when they were larger than 15µL, simply because of their weight. The precise construction of the surfaces was such that freshly-cleaved mica, 0.05mm thick, was cut into various shapes and placed on a soft paraffin wax surface. The water contact angle, the advancing contact angles, the receding contact angles and the contact-angle hysteresis of mica, wax and low-density polyethylene were determined (table 13). The mica surface comprised mainly the hydrophilic -OH group while the wax and polyethylene surfaces comprised mainly hydrophobic groups ($-CH_2-$, $-CH_3$). The mica, wax and polyethylene were thus strongly hydrophilic and hydrophobic, with water contact-angles of 7°, 109° and 102°, respectively. When a droplet was placed at the contact-line of a smooth mica/wax or mica/polyethylene composite surface the droplet spontaneously spread from the hydrophobic (wax, polyethylene) region to the hydrophilic (mica) region. When a droplet was placed on a hydrophilic surface of limited area, the droplet could spontaneously spread on the surface, among hydrophilic or more-hydrophilic regions. In all cases, the water droplets were usually spherical. Shape-gradient mica/wax and mica/polyethylene composite surfaces were designed as paths for droplets and these paths included a start, a transportation region and a storage region. The gradient start was 0.6mm wide, the transportation region was 17mm long and 3mm wide and the storage area was 7mm wide and 7mm long. The gradient angle was 8°. When a droplet was placed at the gradient start, it counter-intuitively spread immediately from the wax or polyethylene surfaces to the mica surface. It then elongated along the transportation region. The elongated droplet finally shrank from the transportation region to the storage region. The droplet running covered 1.7cm within 0.17 and 0.25s, and the average velocities could be as high as 10.0 and 6.8cm/s, respectively. The water droplets could run uphill, under no external force, on shape-gradient composite surfaces which were inclined at 25° to the horizontal. The process included a spreading stage, an elongating stage and a shrinking stage. The run-times of the 1.7cm length were 0.45 and 0.75s, and the average velocities were 3.8 and 2.3cm/s, respectively. When the inclination was greater than 28°, there was no uphill droplet motion on the shape-gradient composite surfaces. Four geometrical mica/wax composite surfaces were considered: trapezoid, triangular, rectangular and meniscus (tables 12 to 14). On the trapezium composite surface, water droplets could run horizontally at high (10cm/s) velocities, and uphill at 3.8cm/s, over a distance of 24.5mm. On a triangular composite surface, the area of the gradient-start was small and the spreading was consequently much slower than that on the trapezium composite surface. Droplets could here run horizontally at 6.7cm/s, and uphill at 2.3cm/s, over a distance of 24.0mm. On a rectangular surface the spatial limitation of the transportation area was great and this markedly affected the elongation stage of running. A droplet could thus run horizontally only at a velocity of 2.4cm/s over a distance of 26.0mm. At an angle of 25°, droplet motion was impossible. On the meniscus surface, with a continuous increase in gradient angle, the running was about the same as that on the trapezoid surface. Unlike the running on the latter surface, the droplets could run at higher (13.1cm/s) velocities horizontally, and uphill at 6.4cm/s. The maximum horizontal distance was however 16.3mm, and the uphill distance was 14.0mm, because of the continuous increase in gradient. It was concluded that the trapezium surface was the best for driving water-droplet motion at high velocities over the greatest distance. The surface gradient and the droplet volume were important (figures 9 to 11). In order to determine the relationship between gradient and droplet motion, mica/wax composite surfaces which were 0.6mm wide at the start and 40mm long were inclined at 0°, 8°, 14°, 16°, 18.5°, 22°, 30°, 35°, 40° or 45°. As the angle was increased from 0° to 45°, the droplet velocity increased from 2.4 to 15.2cm/s while the maximum distance of travel decreased from 26 to 4mm. The effect of droplet size on maximum travel was such that, on horizontal mica/wax and mica/polyethylene surfaces, the droplets could run from the gradient start to the storage region even when they were as large as 50μL. With increasing droplet volume, the average velocity decreased from 10.0 and 6.8cm/s to 1.8 and 0.6cm/s, respectively. On surfaces which were inclined at 25°, a droplet could not run uphill when its size was greater than 15μL.

Table 12. Sessile contact angles and contact-angle hysteresis of mica, wax and low-density polyethylene

Material	Contact-Angle(°)	Advancing CA (°)	Receding CA(°)	Hysteresis(°)
mica	7	11	6	5
wax	109	111	100	11
polyethylene	102	104	68	36

The driving force in all cases arose from the difference in interfacial tension between the droplet and the solid material and a resistive force arose from contact-angle hysteresis on the solid surface. The imbalance arising from the great wettability difference of the surface materials was calculated to be about 97.2 and 87.4mN/m for the mica/wax and mica/polyethylene surfaces, respectively. The resistive force was about 12.3 and 44.9mN/m for the mica/wax and mica/polyethylene surfaces, respectively. The driving force exceeded the resistive force by 84.9 and 42.5mN/m on the mica/wax and mica/polyethylene surfaces. The droplet could however not attain the balanced water contact-angle (7°) at the gradient start because of the confined space of the gradient start (a width of 0.6mm). The droplet therefore had a temporary imbalance contact-angle of 40°. The force was then about 16.5mN/m on both the mica/wax and mica/polyethylene surfaces. Driven by this force, the droplet then immediately elongated itself along the transportation region so as to attain the balance water contact-angle of 7° on the mica surface. On surfaces at 25° to the horizontal, the elongation stage was prevented by a gravitational force of 7.5mN/m. In most experiments, the driving force exceeded the droplet gravitational force by 9.0mN/m, so the droplet could elongate uphill. The differing velocities on a given mica/wax surface, at differing angles, were due to the elongation stage. For a given mica/wax or mica/polyethylene surface at an angle of 8°, the average velocity of a water droplet on the mica/polyethylene surface was lower, both horizontally and uphill, than that on a mica/wax surface. During the spreading stage, the driving force on the mica/polyethylene surface decreased to 42.5 from 84.9mN/m because of the lower water contact-angle (102° rather than 110°) and the higher contact-angle hysteresis (36° rather than 11°). The decrease in velocity was therefore due to the spreading stage while the elongation and the shrinking stages played no role. The difference in total was due mainly to the spreading stage while the elongation and shrinking stages were relatively constant.

Table 13. Average velocities and maximum distances of water droplets on mica/wax composite surfaces

Surface Geometry	Orientation	Velocity(cm/s)	Distance(mm)
trapezoid	horizontal	10	24.5
trapezoid	25°	3.8	24.5
triangular	horizontal	6.7	24.0
triangular	25°	2.3	24.0
rectangular	horizontal	2.4	26
rectangular	25°	0	0
meniscus	horizontal	13.1	16.3
meniscus	25°	6.4	14.0

On the mica/polyethylene surface, the end-times of the spreading stage were 0.10s and 0.40s for the horizontal and uphill conditions, respectively. These were longer than for the mica/wax surface (0.05s and 0.15s). The end-times of the elongation and shrinkage were approximately equal (0.12s, 0.15s) for the horizontal and uphill conditions (0.30s, 0.35s). The decrease in the velocity was thus due mainly to the spreading stage while the elongation and the shrinkage stages had no effect. For mica/wax surfaces at differing angles, the space limitations of the transportation region were different: a greater angle implied a sharper decrease in the transportation region. With space limitation, running could be accelerated because the surface energy also decreased sharply.

A seminal study was made[115] of liquid droplets which moved against gravity when placed on a vertically vibrated inclined plate which was partially wetted by the drop. The frequency of the vibrations ranged from 30 to 200Hz and, at above a threshold vibration-imposed acceleration, the droplets exhibited upward motion. This counter-intuitive observation was attributed to deformations of the droplet which resulted from an up or down symmetry-breaking which was due to the presence of the substrate. This could be used to move a droplet along an arbitrary path. Under normal circumstances, a liquid droplet on an incline will slide downward due to gravity. One exception occurs when the droplet is pinned by contact-angle hysteresis.

Table 14. Driving force of water droplets on surfaces

Surface	Orientation	Stage	Velocity(cm/s)	Force(mN/m)
mica/wax	horizontal	spreading	10.0	84.9
mica/wax	horizontal	elongation	10.0	16.5
mica/wax	horizontal	shrinkage	10.0	10.0
mica/wax	uphill	spreading	3.8	84.9
mica/wax	uphill	elongation	3.8	9.0
mica/wax	uphill	shrinkage	3.8	2.5
mica/polyethylene	horizontal	spreading	6.8	42.5
mica/polyethylene	horizontal	elongation	6.8	16.5
mica/polyethylene	horizontal	shrinkage	6.8	10.0
mica/polyethylene	uphill	spreading	2.3	42.5
mica/polyethylene	uphill	elongation	2.3	9.0
mica/polyethylene	uphill	shrinkage	2.3	2.5

Such hysteresis is reduced by vertical vibration, so it was logical to anticipate that vigorous shaking would loosen the droplet and thus allow it to slide. The novel discovery was that sufficiently strong harmonic shaking in the vertical direction would always a droplet to climb up a slope, regardless of other factor. The upward force was deemed to involve a combination of a broken symmetry which was caused by the slope of the substrate with respect to the applied acceleration, and by a non-linear frictional force between the droplet and the substrate. During downward acceleration, the droplet became taller and therefore more susceptible to lateral forces. The maximum value of the contact-angle on the upper side was greater than the maximum value on the lower side, and the droplet therefore experienced a nett upward force. In the case of a purely linear frictional force, the nett force on the droplet would of course average to zero over a given vibrational period. Some non-linearity therefore had to exist in the interaction between the droplet and the substrate.

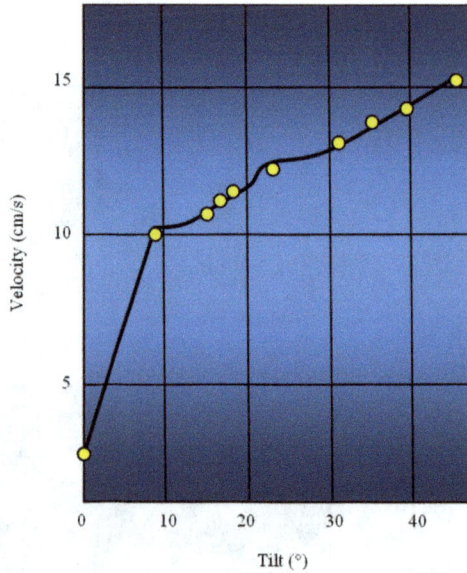

Figure 9. Average velocities of water droplets running on surfaces at various angles

In an experimental study, a droplet of glycerol-water solution with a volume of between 0.5 and 20μL was deposited on a plexiglas substrate which was inclined to the horizontal by an angle of up to 85°. The associated sessile drop was between 1 and 3mm in diameter and, in the absence of shaking, was fixed in place. The substrate was then oscillated vertically, using accelerations of up to 50g and frequencies of between 30 and 200Hz. The lateral acceleration was less than 3% of the vertical acceleration. The kinematic viscosity of the solutions which were used ranged from 31 to 55mm²/s. In the case of low viscosities, the droplet could break up prior to climbing. In the case of high viscosities, the droplets moved more slowly and their dynamics were more difficult to monitor. The surface tension was 0.066N/m and the density at 20C ranged from 1190kg/m³ when the viscosity was 31mm²/s to 1210kg/m³ when the viscosity was 55mm²/s. The advancing and receding contact-angles were 77° and 44°, respectively. The amplitude and frequency of the imposed vibrations produced a rocking of the droplets. When this rocking motion was sufficiency large, the contact-line began to unpin. The resultant mean motion of the droplet depended upon the acceleration and frequency (figure 12). The droplet could move (slide) down the substrate, could remain stationary or could move up (climb) the substrate. Similar phase diagrams were found for various droplet-volumes, viscosities and slopes.

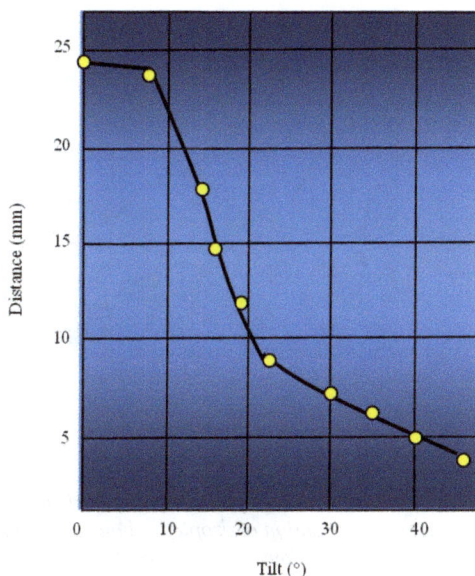

Figure 10. Maximum distances travelled by water droplets running on surfaces at various angles

The boundaries of the various regimes changed as the parameters were varied, but the overall appearance of the phase diagram was unchanged. The manipulation of sessile droplets is of interest for microfluidic purposes, and droplets can clearly be moved by independently varying the phase and amplitude of the vertical and horizontal vibrations for each axis. Spontaneous droplet motion due to gravity, wettability gradients, asymmetrical vibration and chemisorption was already known, but this then-new mechanism would work for uniform substrates, zero mean-forcing and zero externally-imposed gradients.

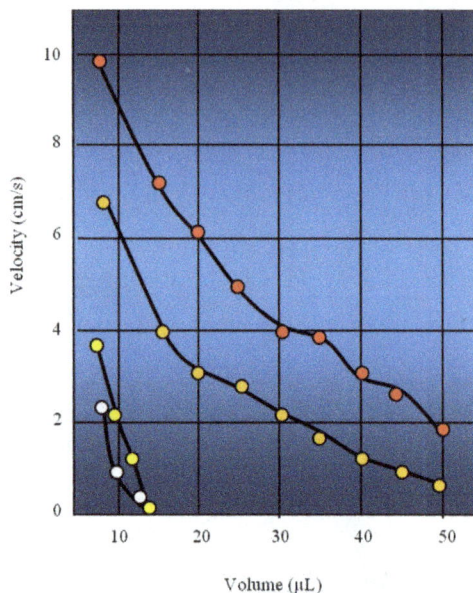

Figure 11. Average velocities of water droplets, running horizontally or uphill (25°) on various composite surfaces, as a function of droplet-volume. Red: mica/wax (horizontal), orange: mica/LDPE (horizontal), yellow: mica/wax (uphill), white: mica/LDPE (uphill)

The technique of photovoltaic charge lithography is habitually used to print surface charges, from illuminated iron-doped lithium niobate crystal stamps, onto passive dielectrics. It has been proposed that the method might also be useful for droplet-manipulation via electrowetting, as this would not require photosensitive materials. Incoherent illumination in air was first used to study the effects of light exposure upon electrowetting and dielectrophoretic droplet-attraction, and this confirmed the ability of photovoltaic charge lithography to manipulate droplets on various passive dielectric substrates. It was shown[116] that wetting properties could be tailored via the exposure time used for charging. Droplets could be attracted to printed charges due to dielectrophoresis.

Single and multiple consecutive droplet motion was possible, including uphill on a tilted substrate. The method was applied to droplet transport on PTFE, and droplets were placed in a shifted position with respect to the charged area. The droplet was immediately attracted to the edge of a printed charge, even if the substrate was tilted, and the droplets could move uphill. The acceleration of droplet-motion was proportional to the area of the printed region. The method therefore permitted the use of light-induced virtual electrodes of any shape, and was not limited by photosensitive material or fabrication process.

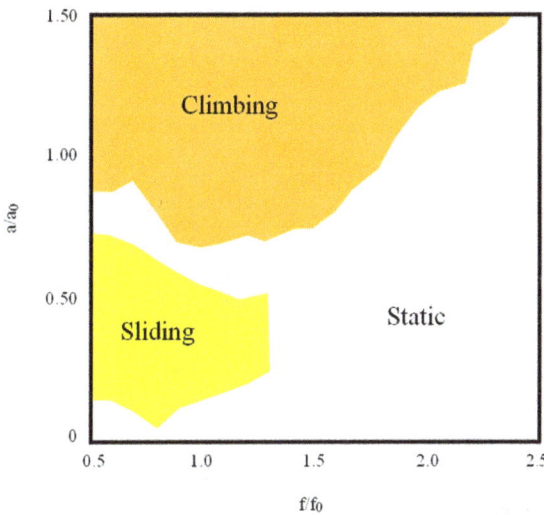

Figure 12. Frequency-acceleration phase diagram for droplets with a volume of 5 μL, on an incline of 45° and having a kinematic viscosity of 31 mm²/s, with normalization factors of f_0 = 50.77Hz and a_0 = 174m/s²

A photo-responsive surface which combined the characteristics of photothermal Fe_3O_4 nanoparticles and low-hysteresis lubricant-infused organogel was studied[117]. A photothermally-induced dynamic temperature gradient formed rapidly at the point of near-infrared irradiation of a suitable Fe_3O_4 nanoparticle content. Droplets of water, glycerol, ethylene glycol, propylene glycol and ethanol could be transported rapidly along extended trajectories at a controllable velocity, and could even run uphill. There was a synergism between asymmetrical droplet-deformation and internal Marangoni flows which provided an unusual driving force for droplet transport. Photothermal organogel surfaces could generate large driving forces for droplet motion under near-infrared irradiation. In a typical situation, a large (circa 7μL) water droplet could be propelled uphill by such irradiation at an average speed of some 0.48mm/s on a 10° incline. A small (circa 2μL) droplet moved downwards due to gravity and, when they merged, the new droplet continued to move upwards at an average velocity of about 0.33mm/s. A 5μL water droplet could move uphill on a 14.7° incline. A 3μL light-actuated water droplet could also deform so as to capture a 0.97mm silica microsphere and carry it onward. A glycerol droplet could also sweep up aggregates of less-than-90μm cellulose particles for self-cleaning purposes. When a 5μL propylene glycerol droplet on a photothermal organogel surface was subjected to near-infrared radiation, one side of the droplet shrank and the droplet moved to the other side due to the imbalance in surface tension between the two sides of the droplet. It was proposed that the resultant temperature gradient acted on the receding side of the droplet and the temperature imbalance between advancing and receding droplet edges could be maintained during long-term transport. The flow within light-actuated droplets could be deduced by monitoring 150μm beads within 15μL water droplets when irradiated on one side. This confirmed that the temperature-field induced Marangoni flow at the liquid/gas interface of the droplet.

Again inspired by living structures, ordered magnetic microcilia arrays have been generated by using a magnetically-induced self-growth method. Fluoride-free superhydrophobic magnetic microcilia arrays were introduced[118] into hexadecyltrimethoxysilane-modified SiO_2 solution. Exploiting the Cassie state, the micro/nano hierarchical structure exhibited surface superhydrophobicity. Due to bending of the cilia, superhydrophobic magnetic microcilia array surfaces enabled droplet manipulation in a magnetic field. The superhydrophobic magnetic microcilia array surfaces had a micro/nano rough structure, and included a large number of nano-SiO_2 features. Modified SiO_2 was uniformly scattered throughout the surface. Such a magnetic microcilia array surface was able to capture and release droplets vertically. The arrays could also promote the horizontal transport of droplets, plus antigravity transport up an 8° incline. These two possibilities permitted the 3-dimensional manipulation of droplets across obstacles. In addition to transporting water droplets, the surface could promote droplet-mixing. Due to the ultra-low adhesion of liquids to superhydrophobic surfaces, it is generally difficult for objects to move uphill.

General Vibratory Phenomena

So-called powder 'droplets', hereinafter referred to as clumps, were observed[119] to form when a horizontal flat plate which was covered with a monolayer of fine particles was repeatedly tapped. It was deduced that there was a marked analogy between the equations which governed this phenomenon, and the behaviour of wetting liquids. Those equations included an analogy of the Laplace law and of the surface-tension parameter. This in turn led to the material exhibiting instabilities of Rayleigh-Taylor type. Most previous work had considered the physics of granular materials which comprised large (>100μm) solid particles, or the movement of much smaller particles in a vacuum. The dynamic interaction of fine powders with gases or liquids is however pivotal to the behaviour of various particulate chemicals. A particular situation which was considered involved the patterns which were by tapping at a low rate from below a flat container that was half-filled with a deep bed of silica particles of the order of 10μm in size. The fineness of the powder particles meant that there were appreciable air-grain interactions, because the free-fall velocity of the small particles was of the same order-of-magnitude as the forced velocity of the particles that was due to the external perturbation. Tapping at a low rate simplified the analysis because it permitted the system to relax between successive perturbations and thus avoided the complication of couplings to vibrational modes. Under the chosen conditions, a quasi-periodic steady-state corrugated pattern spread out. This had a characteristic wavelength which was proportional to the tap amplitude. The instability was deemed to occur because particles were counter-intuitively more easily ejected by air blown from the tops rather than from the sides of heaps. This was termed a 'volcano effect'. The resultant pattern was characterised in terms of a cut-off length which resulted from a competition between downstream avalanches and up-coming particles which were ejected by the trapped air flow. When a thin slice of fine (10 to 50μm) powder spread on a flat plate was gently and repeatedly tapped at constant intensity onto the plate, there occurred the formation of separate rounded and conical clumps spread evenly over the plate. The resultant pattern markedly resembled the Rayleigh-Taylor instability which would be observed if a glass plate covered with a thin film of liquid were to be inverted. This was concluded to result from the underlying similarity between the equations which govern the wetting of liquids and of the behavior of powder piles which interact with a surrounding fluid. That is, surface tension was 'equated' to convective forces, the droplet radius was equated to the clump height and the law of droplet equilibrium was equated to the law of clump equilibrium. By using this analogy it was possible to exploit the traditional demonstration of Rayleigh-Taylor instability for wetting liquids. The standard analysis models the evolution of an infinitesimal sinusoidal distortion of an initially flat surface. This leads to a wavelength (mean clump separation) of the form,

$$L = \propto (\gamma/\rho g)^{3/4}$$

where γ is an ersatz 'capillary force' and ρ is the particle density. This showed that the underlying behaviour of the powder parallels the familiar Rayleigh-Taylor instability. Pursuing the analogy made it possible to predict that, if two powder clumps of unequal sizes were adjacent to one another, the smaller one would be absorbed by the larger one.

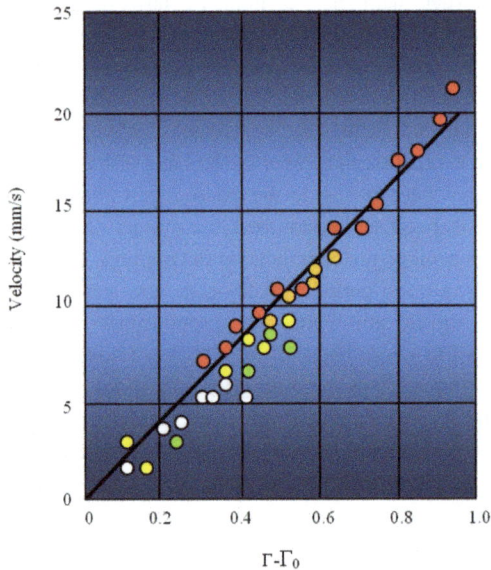

Figure 13. Clump velocity as a function of the difference in the dimensionless acceleration and its critical value for various frequencies. Red: 10Hz, orange: 15Hz, yellow: 20Hz, green: 25Hz, white: 30Hz

But the really relevant question in the present work is what happens if the supporting surface is not horizontal. The physics of vertically-vibrated granular materials is also of interest in fields ranging from mining to pharmaceuticals. The climbing of fine powder particles was again shown[120] to result from interstitial effects. Clumps formed when an inclined surface, which was originally covered in fine powder, was vertically vibrated. The velocities of the clumps were essentially independent of their size, and they could remain stationary at the top of a convex surface, due to the effect of the clump dynamics. When subjected to vibration with sufficiently high acceleration, a layer of grains repeated loses contact and collides with the container, leading to heaping and convection. When the container is evacuated, heaping is no longer observed but patterns such as surface waves are observed. The latter resulted from a transition in the collision time. Clumps were found to form when a horizontal plate, covered with a monolayer of fine particles was tapped. There was an analogy between the equations which governed this phenomenon, and wetting with liquids. The Rayleigh-Taylor instability in particular was deemed to be responsible for selecting the clump-size. The uphill motion of powder clumps on vertically vibrated inclined surfaces shared some characteristics of the wetting of a fluid on a solid surface, but was essentially different.

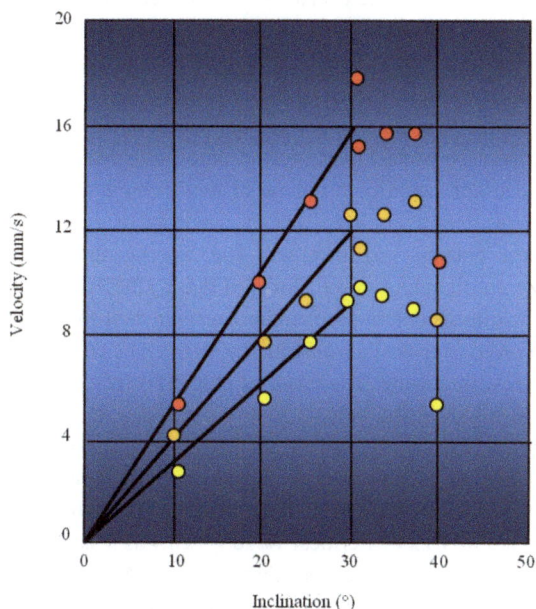

Figure 14. Clump velocity as a function of inclination angle for various values of dimensionless acceleration. Red: 1.6, orange: 1.5, yellow: 1.4

On horizontal surfaces, the powder clumps had symmetrical shapes and, apart from a small random shift, they did not move. In the case of inclined surfaces, the clumps moved uphill. In regions of high inclination, granular fingers appeared but, as the finger-tips moved to surfaces of lower inclination, the fingers broke up into clumps. At the top of a convex surface, a clump of well-defined size could remain stable. The main reason for this phenomenon was the drag of interstitial air on the porous clumps.

In an experimental study, an electromechanical vibrator was used to impose sinusoidal movements. The instability threshold of climbing clumps was measured on symmetrical cones with various angles. The vibration frequency was varied from 10 to 80Hz, and step changes were made in the vibration amplitude. The dimensionless acceleration, Γ, $(4\pi^2 f^2 A/g)$ and f were used as control parameters, where f was the frequency and A was the amplitude. The results showed that an initially horizontal thin layer spontaneously broke up into small clumps when Γ was larger than a critical value. The free surface of a clump was inclined such as to approach the angle-of-repose at the contact line. Within the clumps, in the steady state, particle avalanches at the free surface were are balanced by ascending bulk transport. On a horizontal surface, the clumps had a symmetrical conical shape and the contact-lines were essentially circular. Small clumps were unstable and slowly drifted until they were absorbed by the largest clumps. When the vibrated surface was inclined, the clumps remained almost conical. Because the conical clumps were cut by the inclined surface, the contact lines became almost elliptical. The surface avalanches became asymmetrical and could both oppose or assist clump-climb. The uphill motion was almost independent of clump-size, and the intensity of avalanches decreased with inclination while the clump-speed increased. The experimental evidence showed that, during a cycle, the entire clump moved forward due to the effect of air-flow. At zero inclination, and for dimensionless accelerations greater than the critical value, clump-formation was observed over the entire frequency range. The critical acceleration was an increasing function of frequency. The onset of clump-formation and climb was almost independent of the angle. In the case of inclined surfaces, climbing clumps appeared when Γ attained a critical value. Above a maximum frequency which was a sensitive function of the inclination, clumps were no longer observed. This critical frequency was attributed to a progressive decrease in air-drag influence as the frequency increased. For each inclination, there was another critical value which was a slowly increasing function of frequency and where clumps were no longer stable. Above this critical value, the climbing clumps shrank by ejecting particles until they disappeared. Convection required that clump-plate collisions should remain synchronized to the external excitation but, if the amount of energy which was injected into the layer during the collisions was too great, it could destroy the packing of clumps. This energy was of the order of $g^2(\Gamma-1)^2/f^2$

and, as compared with the potential energy of the packing, became important at low frequencies. On inclined surfaces, clumps of various size coexisted and this made it possible to measure their velocity as a function of size while the experimental parameters were held constant. Size-selection did not occur but the clump velocity was nearly independent of droplet size and clearly selected. The clump speed was a function of the frequency and the acceleration of the vibrating surface. As the frequency was increased at a given inclination, the interval in Γ which was required for the existence of climbing clumps decreased (figures 13 and 14). Most of the experiments were carried out for an rms surface roughness of about 100μm. This was of the same order as the particle-diameter. A high roughness could affect clump stability when the inclination approached the characteristic friction-angle of the surface. As the surface roughness decreased, the limiting angle at which clumps disappeared then decreased. When smooth glass spherical particles were studied, the domain-of-existence of clump-climbing was sharply reduced. An approximation to the overall behaviour proposed that, during each cycle, the clumps lost contact with the plate when the latter's acceleration was smaller than -g. For a sufficiently high Γ-value, the clumps underwent a flight which was of the order of the excitation period, while the relative plate-clump distance was almost equal to the maximum amplitude of the plate. In the earliest stage of clump take-off, the effect of interstitial air flowing through the layer was to limit the relative clump-plate velocity to the dissipation-rate due to air-friction. When the gap between the plate and the clump bottom was of the order of the particle diameter, the friction which was due to interstitial air-flow markedly decreased because the flow through the clump-plate gap became more favourable and the clump moved freely under the influence of gravity. Clumps therefore formed if the time required to open a sufficiently large gap was shorter than the clump flight-time in the drag regime. A natural scaling factor for the velocity of clumps over the inclined surface followed when it was assumed that, during the time-of-interaction of the interstitial flow with the granular layer, clumps accelerated towards the plate at the plate acceleration. That is, in the early stage of take-off the clumps were sucked by the plate and moved along the normal to the vibrating plane. The lateral droplet acceleration was approximately gsinα, where α was the inclination.

Experiments showed[121] that the motion of a droplet on a substrate could be produced by symmetrical vibrations which were at a certain angle to the substrate. This motion was due to the averaged Young's capillary forces which were induced by the detailed motion of the centre-of-mass of the droplet during its rocking motion. When the substrate was inclined, the vertical vibrations led to spectacular climbing. The shape-transitions resembled those observed in gravity-driven sliding droplets but the local dynamics were actually very different. The substrate vibrations caused constant depinning of the contact-line even when the contact-angles lay between the receding and advancing values. The depinning occurred at every point around the drop during sliding and climbing but some parts remained stuck in the case of static drops.

Later studies built upon the seminal work of Brunet *et al.*, and demonstrated[122] that drops of liquid which are placed on vertically oscillating inclined planes are able to climb uphill. A 2-dimensional shallow-water model was used to incorporate surface tension and inertia effects and qualitatively reproduce the main features of the experimental observations. The motion of a drop was controlled by the interaction of a swaying (odd) mode, driven by in-plane acceleration, and a spreading (even) mode which was driven by cross-plane acceleration. It was necessary that both modes be operating if the droplet was to climb uphill. The effect was greatest when the modes were in phase. A liquid droplet on a stationary incline either slides downhill or remains stationary. The latter occurs only if there is a sufficient difference between the advancing and receding contact angles. The assumption made was that, if the plane was vibrated, a sliding droplet would continue to slide and a stationary droplet would either remain stationary (under weak vibration) or would start to slide. Further experiments[123] showed that the average velocity of the droplet could be 'tuned' to a given value by varying the phase-shift between the two components of the oscillation. Theoretical studies[124] of droplets on vibrating substrates approximated a droplet as being a forced linear oscillator having a frequency equal to that of the droplet's free oscillations, showing that the drift direction was determined by the anharmonic component of the vibration. Interaction of the substrate acceleration, inertia and surface tension is sufficient to explain the rising of a drop on a vibrating plate. In linear analyses, appreciable mean velocities occur only away from the resonant frequency of the droplet. Droplet motion is controlled by weakly non-linear interactions of the swaying and spreading modes, and this applies even when it becomes entirely non-linear. In the absence of friction, droplets slid downwards regardless of other parameters.

It was again demonstrated[125] that vertically-vibrated horizontal ratcheted substrates can produce unidirectional motion and, for inclined substrates, attention was focussed on qualitative features of the counter-intuitive vibration-induced climbing of droplets. These included the effects of weak inertia on their dynamics, criteria for uphill motion in the limit of weak gravitational and vibrational effects and the effect of heterogeneities. It was again asked how the combined effects of gravity, substrate-vibration and substrate heterogeneities affected the dynamics of 2-dimensional droplets. Long-wavelength models were used and the dynamics were assumed to be quasi-static. The technique of matched asymptotic expansions was used to derive a set of integro-differential equations for the evolution of the two droplet fronts in the limit of long oscillation periods. In the limit of low-frequency vibrations, an expression was obtained for the mean drift velocity.

This calculation provided a criterion for climb such that, upon strongly vibrating the substrate, uphill motion could occur. Other results implied that climb could not be observed in the low-frequency limit. Also considered was the inclusion of weak inertial effects, and large changes in the high-frequency limit, such as a reduced average speed and the existence of a secondary climb regime. This secondary regime was also present in the inertia-free case, but droplets had to suffer strong vibration at high frequencies which could make them liable to instability. There were so-called static-parameter regimes within which no nett transport occurred. These normally existed at higher frequencies and were attributed to inertial effects. In order to explain the static regimes, it was necessary to include substrate heterogeneities which led to angle hysteresis. Although the model did not assume hysteresis, it could exhibit hysteresis-like effects if the heterogeneities varied over small length-scales. Substrate heterogeneities markedly affected the underlying dynamics at both low and high frequencies. It was in fact demonstrated that heterogeneities could both increase or impair droplet transport, either uphill or downhill, by using ratcheted substrates. This could occur only if the vibration was such that it broke the to-and-fro symmetry of lateral motion. An alternative mechanism could break the symmetry, such as heterogeneities of the substrate.

In order to investigate further the counter-intuitive uphill movement of a liquid droplet on an inclined plane under strong vertical oscillation, a model was proposed[126] in which liquid inertia and viscosity were taken to be negligible. The motion of the droplet was then dominated by the acceleration due to the oscillation of the plate, by gravity and by surface tension. The motion of the droplet could be separated into a spreading-mode and a swaying-mode. Assuming a linear contact-line law, the maximum rising velocity occurred when these modes were in phase. With or without contact-angle hysteresis, the droplet could climb uphill. For certain contact-line laws, the droplet motion matched experimental results. When the modes were out-of-phase, and there was no contact-angle hysteresis, the assumption of hysteresis could force them into phase. This then increased the rise velocity of the droplet and could sometimes cause a sliding droplet to climb. A small hysteresis-interval had little effect upon the overall movement of the droplet but, with a large enough hysteresis-interval, the droplet could climb uphill rather than sliding down. For sufficiently large hysteresis-intervals, the droplet oscillated, but the contact-line was completely pinned. The non-linear spreading-mode was determined by plotting the rate-of-change of distance between the leading and trailing edges of the droplet on the line of symmetry. The non-linear swaying-mode was determined by plotting the velocity of the mid-point of those edges: that is, the centre of the droplet. It was assumed that the slope of the free surface was small. When the amplitude of the oscillation was sufficiently small, the droplet motion could be separated into 2 modes and their non-linear interaction caused the droplet to slide or climb. It could climb uphill for most parameter values, provided that the oscillation-frequency was large enough, with the rise velocity attaining its maximum value when the modes were in phase. The addition of hysteresis could cause a sliding droplet to climb by forcing the modes into phase and producing a larger rise velocity.

The ratchet mechanism of drops climbing a vibrating inclined plate was studied[127] by means of the 3-dimensional numerical simulation of experimental results. The results revealed a significant wetting behaviour of a climbing drop, in that there was a symmetry-breaking of symmetry due to the inclination of the plate with respect to the acceleration. This led to a hysteresis of the wetted area during one period of harmonic vibration. The average wetted area during the downhill stage was larger than that during the uphill stage, and this was responsible for the uphill nett motion of the drop. A novel hydrodynamic model was used to interpret the ratchet mechanism by taking account of the effects of the acceleration and contact-angle hysteresis. This was the first time that the combined effects of vibration, inertia, viscosity, gravity and contact-angle hysteresis had all been taken into account in 3-dimensional numerical simulations. These simulations permitted access to the time-variation of the wetted area of the droplet. It was found that the droplet had a larger wetted area when it slides down than when it moves up. The novel hydrodynamic model showed that the asymmetry of the wetted area was responsible for the uphill nett displacement of the droplet. The model related the nett droplet velocity to the vibration amplitude and frequency, to the contact-angle hysteresis and to the non-harmonic response of the droplet. The presence of gravity rendered the droplet prone to sliding down the inclined plate but, because gravity was much weaker than the acceleration of the plate, it was supposed to exert little effect upon the droplet shape. It was also expected that the gravitational force would not affect the wetted area of the droplet. As a result, the droplet could adjust its own nett velocity in such a way that the gravitational force was eventually balanced by the resultant upward nett viscous force which was exerted by the plate. A non-linear capillary force and a non-linear viscous force could coexist in the ratchet mechanism of nett uphill droplet motion.

The directional motion of a 2-dimensional droplet on an inclined vibrating substrate was studied numerically. The time-dependent droplet profile was analysed[128] by using an orthogonal decomposition method. Two predominant modes of the capillary wave were identified in which the first mode was quasi-harmonic and led to an apparent difference in wetted area between the uphill and downhill stages of substrate vibration. This played a pivotal role in directional motion. The second mode was weak but made a subtle contribution. The two modes qualitatively matched the known swaying and spreading modes. The present decomposition directly revealed the connection between the surface waves. The 2-dimensional model which was used provided simplified but valid data for analysis. Numerical simulation of a droplet climbing on an obliquely vibrating plate showed that the free surface executed a rocking motion which was associated with differing separate capillary wave evolutions during the uphill and downhill stages. During the uphill stage, a capillary wave travelled along the free surface following the inertial force. Mass-transfer was associated with the capillary wave. During the downhill stage, the traveling capillary wave reached the downstream end of the free surface and disappeared. Another crest was generated meanwhile, and grew near to the upstream end of the free surface. The fluid mass was inflated from one part to another under the action of inertial force. The free-surface motion was further decomposed by using a proper orthogonal decomposition method. The first two modes reflected the surface motion at any given time. The first mode corresponded to a standing wave having nodes near to the two ends and to the centre of the free surface. This mode was almost sinusoidal and was a linear response to the plate vibration. The second mode had a phase-shift of $\pi/2$ with respect to the first mode, implying orthogonality. The second mode was expected to exhibit extrema near to the ends of the surface, but wall-effects suppressed the extrema and rendered the energy of the second mode much lower than that of the first one. Both mode-I and mode-II were non-trivial with respect to the capillary-wave evolution. The first two proper orthogonal decomposition modes were similar to the proposed swaying and spreading modes. Other analyses had required higher-order quantities and a special contact-line model while the first two proper orthogonal decomposition modes controlled the motion directly. The proper orthogonal decomposition point of view was thus more reasonable.

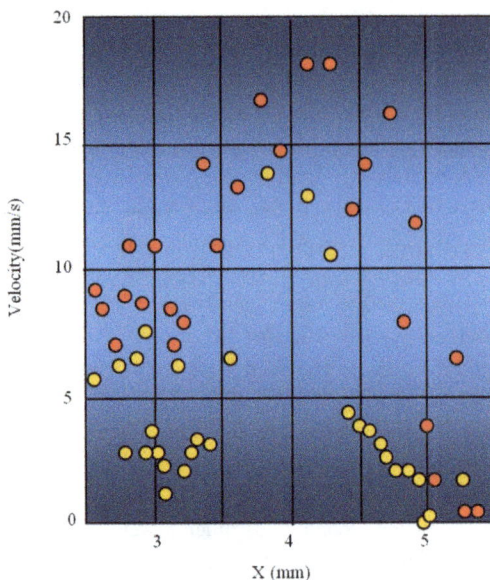

Figure 15. Centre velocity of water droplets of differing sizes on an inclined surface-energy gradient. Red: 1μL, orange: 2μL

Surfaces having a surface-energy gradient were created via the chemical vapour deposition of dodecyltrichlorosilane, and their properties were determined by using sessile drops and atomic force microscopy. The motion of water and ethylene glycol droplets was monitored[129] on horizontal and inclined surfaces. A system free-energy transition was analysed in order to identify the mechanism of droplet self-motion. The height and density of the silane molecular groups governed the surface-energy distribution on the surface.

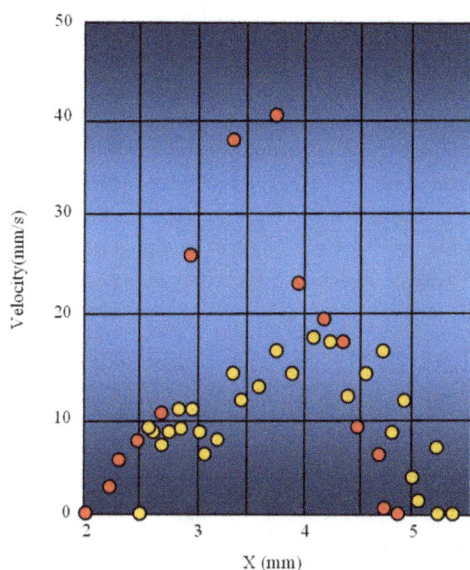

Figure 16. Centre velocity of a water droplet on inclined and horizontal surface-energy gradient. Red: inclined, orange: horizontal

Liquid droplets moved horizontally, or uphill, from hydrophobic to hydrophilic zones on horizontal and inclined surfaces. The motion exhibited an acceleration stage and a creep deceleration stage. The velocity and displacement and the creep frequency were proportional to the droplet-size. The velocity of a 2ml water droplet attained 42mm/s on a horizontal surface and 18mm/s on an inclined surface. Ethylene glycol droplets attained 7mm/s on a horizontal surface. Droplet motion was the result of a transition between interfacial energy, kinetic energy, gravitational potential energy and viscous dissipation, with the interfacial energy released by deformation of the droplet being the main cause of motion. During uphill motion of a water droplet on an inclined surface, the droplet deformed from a spherical shape to a thin film as it moved from a hydrophobic zone to a hydrophilic zone. When compared with the behaviour on a horizontal surface (figures 15 and 16), the velocity of an uphill-moving droplet on an inclined surface fluctuated wildly and gradually increased so as to reach the peak value of 18mm/s at a point away from the initial position. It then decreased until the droplet stopped at a distance of nearly 2.8mm. The velocity during overall motion was lower than that on a horizontal surface, where the maximum velocity was almost 42mm/s and the displacement was nearly 2.9mm. During uphill motion, the static pressure-difference caused by the gravitational force acted against droplet motion. The variation in the centre-of mass of the droplet was greater than that for a horizontal, due to the uphill motion, and this led to an increased negative effect upon uphill motion. This was why the uphill motion was slower on an inclined surface than it was on a horizontal surface. The velocity, and its rate of fluctuation, as well as the displacement for larger droplets were much greater than those for small droplets. Silane molecular groups with a height of nearly 12nm stuck out on silicon surfaces which were silanized with dodecyltrichlorosilane. The height and density of the silane molecular groups determined the surface-energy gradient on the surface.

Most of the present work is concerned with the spontaneous uphill motion of droplets on inclined surfaces but, with regard to the construction of practical devices, it is also necessary to mention the influence of external driving forces. The uphill droplet-transport on a superhydrophobic incline by using an air-flow was demonstrated[130]. When a liquid was delivered continuously via an orifice to a superhydrophobic incline, the increasing weight of the droplets caused the rear contact-angle to decrease to such an extent that the droplet eventually pinched off. Upon making contact with the solid, however, a predominant Cassie wetting-state was restored and this permitted the drop to detach and travel down the surface. This offered a useful means for rapidly generating droplets. The use of air-flow to drive the uphill transport of droplets which formed from water that was delivered continuously via an orifice was studied in detail. Equations were derived, for example, which described the droplet-volume, at the point of detachment at various inclinations, for a range of parameters. Necking and pinching-off before detachment involved a constant time-to-rupture of 0.032s for a neck-thickness of 0.001m, regardless of the inclination angle. The overall approach permitted the rapid uphill translation of droplets with a size of 0.03mL at the rate of 33 droplets per second and water thus being delivered at a flow-rate of 0.1mL/s.

The self-propelled motion of Leidenfrost droplets on ratchet surfaces was further investigated[131] numerically by using a thermal multiphase lattice-Boltzmann model involving a liquid-vapour phase change. An initial study was made of the motion of Leidenfrost droplets on horizontal ratchet surfaces. The motion was due to the asymmetry of the ratchets and to vapour-flow beneath the droplets.

Figure 17. Geometry of ratchet surfaces

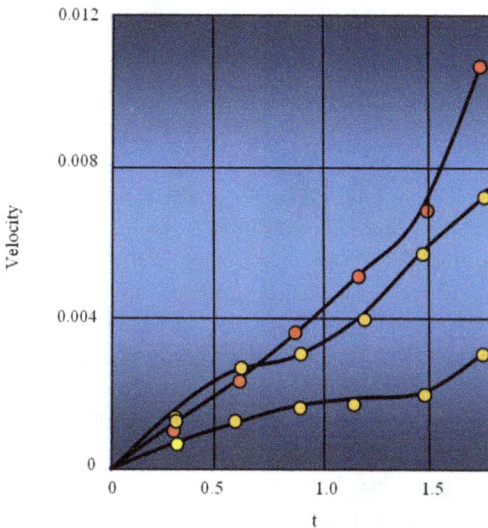

Figure 18. Velocity of Leidenfrost droplets with an initial radius of 40 on various ratchet surfaces. Red: H/L = 1/3, orange: H/L = 5/12, yellow: H/L = 1/4

The droplets moved in the direction of the low-incline side from the ratchet peaks. A critical value of ratchet aspect-ratio was found which corresponded to the maximum droplet velocity. The maximum inclination at which a Leidenfrost droplet could still climb uphill was affected by the initial radius of the droplet. The lattice-Boltzmann method involved a pseudopotential multiphase lattice-Boltzmann model for simulating the density and velocity fields, and a finite-difference solver for the temperature field. The liquid-vapour phase-change was driven by the temperature field via a non-ideal equation of state. The droplet velocity increased when H/L (figure 17) increased from 1/4 to 1/3. It decreased when H/L increased further (figure 18). In the case of a droplet with $R_0 = 35$ on a ratchet surface which was inclined at 4°, the droplet moved downhill in the early stages, due to gravity. The droplet later turned back with an uphill acceleration which was generated by the vapour flow beneath the droplet. The maximum inclination angle at which a Leidenfrost droplet could still climb uphill was related to the initial radius of the droplet

Electrowetting was considered[132] with regard to hydrophobicity and hydrophilicity: Hydrophilic surfaces had a low contact-angle when touching a liquid droplet, while hydrophobic surfaces had a high contact angle. The effect of the surface contact-angle on droplet motion on inclines was investigated analytically. A momentum equation which accounted for the effect of contact-angle change was solved for angles ranging from 30 to 120° and for voltages ranging from 0 to 100V. It was noted that an increase in droplet footprint increased the actuating force for climbing.

The method of electrowetting on dielectrics is the most successful electrical droplet-manipulation method. It is however difficult to ensure repeatability and high speeds on superhydrophobic surfaces. Efficient operation generally requires electrode arrays and control circuits. A novel manipulation method, orbital electrowetting, has been essayed[133]. Due to the asymmetrical electrowetting force which was generated, versatile manipulation was possible and the method produced velocities which were both up to 5 times higher and also enabled uphill motion. In order to permit directional and controllable manipulation, a tapered orbit was used to modify the speed and direction of droplets. Theoretical modelling of the manipulation force was based upon the asymmetrical triple-contact line. The technique could manipulate droplets of de-ionized water, acidic and basic salt-solutions with concentrations of greater than 10wt%, and organic solutions. The velocities could be of the order of 210mm/s, and droplets could be driven up inclines with angles of up to 55°.

The movement of single droplets remained somewhat unclear and the study of eccentric droplets aided clarification of the effect of local energy upon droplets. Based upon existing electrowetting driving-force models and experiment, an electrowetting-force correction factor was proposed[134]. Phase-field simulation was used to model the motion of 10μL droplets, under 250V at 1000Hz, along straight, inclined and circular-arc orbits. In the case of a straight orbit, the spreading radius of a droplet was some 1.09 times the initial radius, the maximum inclination was 9.8° and the maximum correction-force was about 24.7μN at an offset of 297μm. A principal purpose was to study the process by which a droplet overcame gravity. The relative dielectric constant of the droplet was 82, the relative dielectric constant of the dielectric layer was 3.8 and the relative dielectric constant of air was unity. Because a droplet stretched and contracted when spreading, the contact-line between the droplet and controlling electrodes became smaller when spreading and elongating. When the droplet shrank, the contact-line with the electrode became larger and the electrowetting force became larger. The droplet was thus not stable in an inclined orbit under the electrowetting force and adopted the maximum slope-angle which could move. Droplet motion on inclined tracks at angles of 5°, 8°, 9°, 9.5° and 10° was considered. When the slope was less than 9.5°, droplets travelled upward. When the slope was greater than 10°, droplets slid rapidly downwards following oscillation. The maximum slope angle up which droplets could travel was 9.8°. When the inclination-angle of the track was greater than this maximum, the droplets did not slip directly. At an inclination of 10°, the centre-of-mass of the droplet travelled upward along the orbit at the beginning. The droplet then slid down. The droplet vibrated up and down relative to the initial position, and the rises and falls of the centroid position were similar. After a certain time, the position of the centre-of-mass was lower than the initial position and gradually stabilized. The component of the electrowetting force along the orbit direction gradually decreased with descent. Unlike in traditional falling, the electrowetting force perpendicular to the orbit made the droplet oscillate and spread when falling. This caused a slight oscillation when the droplet velocity increased. Energetic-analysis of droplets at various positions showed that there existed a local minimum energy-value near to the centreline of the orbit. This led to droplets being trapped in that area of so-called inherent pinning.

The static and dynamic behaviours of 2-dimensional droplets on inclined heterogeneous substrates were examined[135] from the theoretical point-of-view by using an evolution equation for the droplet-thickness which was based upon the long-wave approximation to the Stokes equations in the presence of slip. Perturbation methods furnished evolution equations for the positions of the two moving fronts, assuming quasi-static dynamics, and these were checked by comparison with numerical solutions in the case of stick-slip dynamics, substrate-induced hysteresis and uphill motion of droplets under strong chemical gradients. There was again a critical inclination above which the droplet could not be supported. Analytical expressions could be obtained for various relevant parameters in the limit of small contact-angles and moderate gravity. A first attempt was made to understand the dynamic behaviour of a droplet on an inclined substrate without imposing the assumption of contact-angle hysteresis. Hysteresis was instead allowed to arise naturally due to substrate heterogeneities. By assuming the occurrence of inertia-less quasi-static dynamics and slip, the evolution equations were derived for the location of two moving fronts via matched asymptotic analysis. It was found that, without assuming hysteresis, droplets could not be kept in equilibrium on perfect substrates. An analytical estimate was deduced, for the rate of descent, which was valid for small gravitational effects and inclinations. When the substrate was heterogeneous, a droplet could attain equilibrium on the substrate. In the presence of a linear chemical gradient which opposed downhill motion, a droplet attained equilibrium for realistic parameter-values. It could also exhibit uphill motion. The analysis was applicable to models which assumed the presence of a film in order to relax the stress-singularity of a moving contact-line. Many qualitative features could be reflected by a 2-dimensional model, but quantitative comparisons with experimental data could not be made. On the other hand, most studies involved large contact-angles, and these were beyond the applicability of long-wave theory. Heterogeneities could assist uphill motion, as when sufficiently high wettability gradients could move droplets against gravity.

A numerical study was made[136] of the dynamics of uphill motion of a sessile droplet, as caused by inducing asymmetrical electrocapillarity. A droplet of de-ionized water in air was used to represent uphill motion under electrostatic forces. The equilibrium contact-angle of the droplet was assumed to be 90°. The density, viscosity and surface tension were assumed to be 1000kg/m^3, 0.00089Pas and 0.072N/m. Neither slip nor penetration occurred at the surface upon which the drop rested. The droplet was placed in a 2-dimensional rectangular container with a height which was equal to twice the height of the droplet and a length which was equal to 6 times the diameter of the droplet. The boundaries were considered to permit neither slip nor penetration. For the purpose of propulsion, a droplet was placed asymmetrically over the electrode-strip array. This had a width which equalled the droplet radius. The droplet was placed over 3 electrodes, with the upper one electrified and the others grounded. A coupled electro-hydrodynamic model was proposed which took account of conservative body and surface forces plus electrostatic effects. The interaction between gravity and electrostatic force was affected by the voltage, the inclination and the droplet-volume. The voltage acting on the sessile droplet caused an internal circulation which, under increasing electrostatic influence, overcame gravity and pulled the droplet uphill. The droplet-volume played an important role in accommodating internal circulations and the resultant climb. The action of the electrostatic force was different on different inclines, with rolling-down occurring at high angles and uphill climb at low angles. The velocity-field within the droplet revealed the predominance of sliding over rolling during uphill motion. The variation in the velocity of the centre-of-mass of the droplet as a function of displacement and time was analysed with regard to the voltage, inclination and droplet-size. At high voltages, the maximum velocity occurred at the end of the activating electrode. The velocity variations on various inclinations revealed the interplay between electrostatic force and gravitational force. The inertia of the bulk fluid which constituted large droplets affected the uphill velocity.

Materials Research Forum LLC

https://doi.org/10.21741/9781644903896

A 2-dimensional inviscid irrotational model was developed[137] for the velocity of contact-lines as a function of contact-angle. Asymptotic analysis was used to show that, for vibrational forces of sufficiently small amplitude, droplet motion could be separated into odd and even modes. Weakly non-linear interaction between these modes again determined whether the droplet climbed, or slid down, a plane. In the weakly non-linear limit it was found that, as the static contact-angle (non-wetting limit) approached, the rise velocity of the droplet, averaged over one period of motion, became a highly oscillatory function of the static contact-angle. This was due to a high-frequency mode which was excited by the forcing vibration. The full non-linear moving-boundary problem was solved numerically by using a boundary integral method. This solution was then used to study the effect of contact-angle hysteresis, which could increase the rise velocity of a droplet, unless the latter was so large that it completely fixed the contact-lines. A time-dependent modification of the contact-line law was also studied in an attempt to reproduce the experimentally-observed unsteady contact-line dynamics. In the latter case, the apparent contact-angle was not a single-valued function of the contact-line velocity. Upon adding a degree of lag to the contact-line model, the rise velocity of the droplet was markedly affected and made larger rise velocities possible.

A magnetically-controlled dimple on a slippery surface was developed[138] which permitted directional or even uphill droplet transport due to a synergy between gravitational force and asymmetrical droplet deformation. Experiments demonstrated that the method could be used for stirring micro-droplets and for accelerating the rate-of-mixing by more than one hundred-fold. Such micro-stirring could avoid the uneven local production of precipitates or gas during heterogeneous reaction. The magnetically-controlled dimple on a slippery surface promoted antigravity transport because, when the magnetic field moved upwards, the dimple carried the droplet uphill. It was observed that gravity drove droplet-flow into the dimple due to the inclined edge of the dimple. The shape of a droplet also changed as it rose, producing an upward Laplace force. The asymmetrical deformation caused by the disturbance of dimple movement then promoted uphill movement. For a static droplet on a slope, according to wettability theory, the apparent contact-angle is governed by Young's formula. Two contact-angles could be defined for the two sides of the droplet. Upon applying a dynamic magnetic field on the right side, the droplet deforms under the action of viscous forces and gravity, the contact-angles change, the equilibrium of the contact-angles of the droplet is disrupted and this produces an unbalanced Young's force which then acts on the droplet.

A phase-field model which included contact-line dynamics was used[139] to analyse the changing contact-angles between droplet and solid. An energy-stable discrete-time version of the state equation was considered. The regularity of the solution to the discrete-time state equation, and its continuity and differentiability, were examined. The existence of solutions and first-order optimality conditions was demonstrated. The results were applied to the optimum uphill pushing of a droplet.

The bouncing of droplets on super-repellent surfaces[140] is analogous to specular reflection, so much so as to obeys the law of reflection. The non-specular reflection of droplets from solid surfaces containing a dimple was studied, with the dimple radius being comparable to that of the droplet. The droplet had a translational velocity that could be varied in direction and magnitude by changing the droplet radius, the dimple radius, the impingement-point and the droplet Weber number. Various means for droplet manipulation via non-specular reflection on dimple surfaces were considered. Trapping and shedding could be steered merely by adjusting the dimple wettability. A droplet was trapped when the dimple was superhydrophobic, but was shed when the dimple was less hydrophobic. This was unexpected because it is normal to expect the trapping of droplets by surfaces which offer strong solid/liquid adhesion. A less hydrophobic dimple exerted a larger impulse on the droplet during asymmetrical receding, so that the gain in horizontal momentum endowed the droplet with a higher translational velocity and thus aided shedding. By adjusting the translational velocity and the initial droplet velocity, it was possible to direct a bouncing droplet in any direction. Even antigravity droplet transport was possible in the absence of any external force. This was impossible for conventional super-repellent surfaces. Precise positioning of a droplet, and on-demand droplet coalescence, by varying trapping and shedding on various dimples, was possible. A rebounding droplet always shed towards the dimple centre, with a non-zero translational velocity. By choosing the initial droplet velocity and translational velocity, the trajectory of a reflected droplet could be closely controlled in both direction and bounce speed. The reflection of a droplet on a surface with dimples of comparable radius could thus break all of the symmetries of reflection, unlike specular reflection on nano-/micro-textured surfaces. It was concluded that surface dimples can be used to trap or repel droplets at a chosen velocity. Manipulation via non-specular reflection is energy-efficient, omnidirectional and can maximize both travel distance and velocity.

A method has been proposed for controlling the spontaneous motion of droplets on solid surfaces by using surface-curvature gradients. Molecular dynamics simulations were used[141] to show that droplets on bowl-shaped axisymmetrical surfaces could travel uphill from base to apex and move continuously to the apex at essentially constant velocity. When determining the effect of surface geometry upon droplet motion, the arc-radius and water/solid binding energy were chosen to be 7.5nm and 145K. The dimension, d, ranged from -2nm to 120nm. This was the horizontal distance between the point-of-tangency and the axis of the shape. When d was equal to zero, the point-of-tangency lay on the axis and a quarter-circular arc was the generatrix, thus making the surface a hemisphere. The average velocity of the droplet during motion was a function of d. When d was less than zero, droplets migrated in the downward direction. For intermediate d-values, such as 4nm or 10nm, the direction of motion of the droplet was reversed. The average velocity was about -0.4m/s for various d-values. At d-values greater than 10nm the droplet remained stationary because, as d increased, the size of the driving force was reduced and, with increasing d, the surface became more flat. This increased the extent of the contact-area between the droplet and substrate and caused an increase in the adhesion resistance. When d was greater than a threshold value, the driving force could fail to overcome the resistance. Alternation of the surface structures could be used to control the direction of motion of self-propelled droplets on solid surfaces. In order to guarantee spontaneous motion in the uphill direction, the wettability of the surface had to be suitable. Its surface wettability was varied by changing the water/solid binding-energy. For surface energies ranging from 100K to 200K, upward directed transport of droplets occurred. The average velocity decreased with increasing binding energy, even though the size of the driving force was higher. This was because a high binding-energy favoured fluid adsorption on the wall surface. Increasing the binding-energy caused droplets to spread, thus increasing the water/solid contact area and the adhesion resistance. At surface energies greater than 200K, and under strong attraction by the surface, the droplet underwent a transition from clam-shaped to thin film and remained fixed. For a range of bowl-shaped axisymmetrical surfaces, small droplets on the surface could be caused to move uphill because the induced driving force exceeded the adhesion resistance. For large droplets, due to spatial differences in the vertical direction, the radius of curvature of the bottom meniscus was smaller than that of the upper meniscus. This generated an upward Laplace pressure which could push droplets upward.

The manipulation of droplets on surfaces is complicated by the presence of capillary forces, whose effects become more important when the droplets are close together and tend to cause their coalesce. Triboelectrification has been used to increase the ability to manipulate droplets. During triboelectrification, charges accumulate within the droplets thus permitting droplet manipulation, especially on low-friction superhydrophobic insulating surfaces. The behaviour of water droplets on quartz surfaces was studied[142] following triboelectrification. The droplets acquired an appreciable charge when dropped onto a superhydrophobic glass surface. The charged droplets then exhibited repulsion, uphill motion and rapid long-distance motion. The triboelectric charge on the substrate increased the attraction between substrate and droplets. When a droplet was initially located at a point having a relatively low substrate charge, it could be drawn towards areas having a higher charge by an attractive force of the form, Fcosθ, which was greater that the frictional force, $(mg + F\sin\theta)/\mu$, where θ was the angle between the nett attractive static force and the horizon. When water was continuously dripped, causing the substrate charge to increase, the attractive force between the substrate and droplets permitted them to move uphill. This attractive force remained effective when the two droplets exerted a smaller mutual repulsive force.

About the Author

Dr. Fisher has wide knowledge and experience of the fields of engineering, metallurgy and solid-state physics, beginning with work at Rolls-Royce Aero Engines on turbine-blade research, related to the Concord supersonic passenger-aircraft project, which led to a BSc degree (1971) from the University of Wales. This was followed by theoretical and experimental work on the directional solidification of eutectic alloys having the ultimate aim of developing composite turbine blades. This work led to a doctoral degree (1978) from the Swiss Federal Institute of Technology (Lausanne). He then acted for many years as an editor of various academic journals, in particular *Defect and Diffusion Forum*. In recent years he has specialized in writing monographs which introduce readers to the most rapidly developing ideas in the fields of engineering, metallurgy and solid-state physics. He is co-author of the widely-cited student textbook, *Fundamentals of Solidification*. Google Scholar credits him with 9911 citations and a lifetime h-index of 14.

References

[1] Snoeijer J.H., Brunet P., American Journal of Physics, 80, 2012, 764-771.
https://doi.org/10.1119/1.4726201

[2] Watson G.S., Gellender M., Watson J.A., Biofouling, 30[4] 2014, 427-434.
https://doi.org/10.1080/08927014.2014.880885

[3] Cao M., Li K., Dong Z., Yu C., Yang S., Song C., Liu K., Jiang L., Advanced
Functional Materials. 25, 2015, 4114–4119.
https://doi.org/10.1002/adfm.201501320

[4] Huang S., Song J., Lu Y., Lv C., Zheng H., Liu X., Jin Z., Zhao D., Carmalt C.J.,
Parkin I.P., Journal of Materials Chemistry A, 4, 2016, 13771-13777.
https://doi.org/10.1039/C6TA04908G

[5] Zhang B., Zhu Q., Li Y., Hou B., Chemical Engineering Journal, 352, 2018, 625-633.
https://doi.org/10.1016/j.cej.2018.07.074

[6] Zhang B., Xu W., Zhu Q., Li Y., Hou B., Journal of Colloid and Interface Science,
532, 2018, 201-209. https://doi.org/10.1016/j.jcis.2018.07.136

[7] Nakajima D., Kikuchi T., Natsui S., Suzuki R.O., Applied Surface Science, 440,
2018, 506-513. https://doi.org/10.1016/j.apsusc.2018.01.182

[8] Escobar A.M., Llorca-Isern N., Applied Surface Science, 305, 2014, 774-782.
https://doi.org/10.1016/j.apsusc.2014.03.196

[9] Zheng S., Li C., Fu Q., Hu W., Xiang T., Wang Q., Du M., Liu X., Chen Z., Materials
and Design, 93, 2016, 261-270. https://doi.org/10.1016/j.matdes.2015.12.155

[10] Abdolmaleki M., Allahgholipour G.R., Tahzibi H., Azizian S., Materials Chemistry
and Physics, 313, 2024, 128711.
https://doi.org/10.1016/j.matchemphys.2023.128711

[11] Liu L., Zhao J., Zhang Y., Zhao F., Zhang Y., Journal of Colloid and Interface
Science, 358[1] 2011, 277-283. https://doi.org/10.1016/j.jcis.2011.02.036

[12] Chen Z., Guo, Y., Fang S., Surface and Interface Analysis, 42[1] 2010, 1-6.
https://doi.org/10.1002/sia.3126

[13] Jagdheesh R., García-Ballesteros J.J., Ocaña J.L., Applied Surface Science, 374,
2016, 2-11. https://doi.org/10.1016/j.apsusc.2015.06.104

[14] Liu C., Su F., Liang J., RSC Advances, 4, 2014, 55556-55564.
https://doi.org/10.1039/C4RA09390A

[15] Forooshani H.M., Aliofkhazraei M., Rouhaghdam A.S., Journal of the Taiwan Institute of Chemical Engineers, 72, 2017, 220-235. https://doi.org/10.1016/j.jtice.2017.01.014

[16] Song Y., Wang C., Dong X., Yin K., Zhang F., Xie Z., Chu D., Duan J., Optics and Laser Technology, 102, 2018, 25-31. https://doi.org/10.1016/j.optlastec.2017.12.024

[17] Tong W., Cui L., Qiu R., Yan C., Liu Y., Wang N., Xiong D., Journal of Materials Science and Technology, 89, 2021, 59-67. https://doi.org/10.1016/j.jmst.2021.01.084

[18] Song M., Liu Y., Cui S., Liu L., Yang M., Applied Surface Science, 283, 2013, 19-24. https://doi.org/10.1016/j.apsusc.2013.05.088

[19] Lu S., Chen Y., Xu W., Liu W., Applied Surface Science, 256[20] 2010, 6072-6075. https://doi.org/10.1016/j.apsusc.2010.03.122

[20] Lu Y., Shen Y., Tao J., Wu Z., Chen H., Jia Z., Xu Y., Xie X., Langmuir, 36, 2020, 880-888. https://doi.org/10.1021/acs.langmuir.9b03411

[21] Sheng X., Zhang J., Applied Surface Science, 257[15] 2011, 6811-6816. https://doi.org/10.1016/j.apsusc.2011.03.002

[22] Parin R., Del Col D., Bortolin S., Martucci A., Journal of Physics - Conference Series 745, 2016, 032134. https://doi.org/10.1088/1742-6596/745/3/032134

[23] Lv S., Zhang X., Yang X., Liu X., Yang Z., Zhai Y., Materials Research Express, 9, 2022, 026520. https://doi.org/10.1088/2053-1591/ac433a

[24] Du X.Q., Chen Y., Materials Research Express, 7, 2020, 056405. https://doi.org/10.1088/2053-1591/ab9253

[25] Wang H., Dai D., Wu X., Applied Surface Science, 254[17] 2008, 5599-5601. https://doi.org/10.1016/j.apsusc.2008.03.004

[26] Rezayi T., Entezari M.H., Journal of Colloid and Interface Science, 463, 2016, 37-45. https://doi.org/10.1016/j.jcis.2015.10.029

[27] Xie F., Yang J., Ngo C.V., Results in Physics, 19, 2020, 103606. https://doi.org/10.1016/j.rinp.2020.103606

[28] Ji J., Jiao Y., Song Q., Zhang Y., Liu X., Liu K., Langmuir, 37[17] 2021, 5436-5444. https://doi.org/10.1021/acs.langmuir.1c00911

[29] Liu W., Xu Q., Han J., Chen X., Min Y., Corrosion Science, 110, 2016, 105-113.

https://doi.org/10.1016/j.corsci.2016.04.015

[30] Huang Y., Sarkar D.K., Chen X.G., Materials Letters, 64[24] 2010, 2722-2724. https://doi.org/10.1016/j.matlet.2010.09.010

[31] Guo Z., Fang J., Wang L., Liu W., Thin Solid Films, 515[18] 2007, 7190-7194. https://doi.org/10.1016/j.tsf.2007.02.100

[32] Wan Y., Chen M., Liu W., Shen X., Min Y., Xu Q., Electrochimica Acta, 270, 2018, 310-318. https://doi.org/10.1016/j.electacta.2018.03.060

[33] Shu Y., Lu X., Liang Y., Su W., Gao W., Yao J., Niu Z., Lin Y., Xie Y., Surface and Coatings Technology, 441, 2022, 128514. https://doi.org/10.1016/j.surfcoat.2022.128514

[34] Feng L., Yang M., Shi X., Liu Y., Wang Y., Qiang X., Colloids and Surfaces A, 508, 2016, 39-47. https://doi.org/10.1016/j.colsurfa.2016.08.017

[35] Kuang Y., Jiang F., Zhu T., Wu H., Yang X., Li S., Hu C., Materials Letters, 303, 2021, 130579. https://doi.org/10.1016/j.matlet.2021.130579

[36] Bahrami H.R.T., Ahmadi B., Saffari H., Materials Letters, 189, 2017, 62-65. https://doi.org/10.1016/j.matlet.2016.11.076

[37] Yuan Z., Bin J., Wang X., Peng C., Wang M., Xing S., Xiao J., Zeng J., Xiao X., Fu X., Chen H., Surface and Coatings Technology, 254, 2014, 151-156. https://doi.org/10.1016/j.surfcoat.2014.06.004

[38] Bahrami H.R.T., Ahmadi B., Saffari H., Materials Research Express, 4[5] 2017, 055014. https://doi.org/10.1088/2053-1591/aa6c3b

[39] Chen J., Guo J., Qiu M., Yang J., Huang D., Wang X., Ding Y., Materials Transactions, 59, 2018, 5.

[40] Chaitanya B., Gunjan M.R., Sarangi R., Raj R., Thakur A.D., Materials Chemistry and Physics, 278, 2022, 125667. https://doi.org/10.1016/j.matchemphys.2021.125667

[41] Long J., Fan P., Zhong M., Zhang H., Xie Y., Lin C., Applied Surface Science, 311, 2014, 461-467. https://doi.org/10.1016/j.apsusc.2014.05.090

[42] Shu Y., Lu X., Lu W., Su W., Wu Y., Wei H., Xu D., Liang J., Xie Y., Surface and Coatings Technology, 455, 2023, 129216. https://doi.org/10.1016/j.surfcoat.2022.129216

[43] Shi X., Zhao L., Wang J., Feng L., Journal of Nanoscience and Nanotechnology,

20[10] 2020, 6317-6325. https://doi.org/10.1166/jnn.2020.17891

[44] Zhao Y., Zhang H., Wang W., Yang C., International Journal of Heat and Mass Transfer, 127[C] 2018, 280-288. https://doi.org/10.1016/j.ijheatmasstransfer.2018.07.153

[45] Song J., Xu W., Lu Y., Fan X., Applied Surface Science, 257 [24] 2011, 10910-10916. https://doi.org/10.1016/j.apsusc.2011.07.140

[46] Lv Y., Liu M., Surface Engineering, 35[6] 2019, 542-549. https://doi.org/10.1080/02670844.2018.1433774

[47] Jia C., Zhu J., Zhang L., Coatings, 12, 2022, 442. https://doi.org/10.3390/coatings12040442

[48] Vanithakumari S.C., George R.P., Mudali U.K., Philip J., Transactions of the Indian Institute of Metals, 72[5] 2019, 1133-1143. https://doi.org/10.1007/s12666-019-01586-3

[49] Hassan L.B., Saadi N.S., Karabacak T., International Journal of Advanced Manufacturing Technology, 93, 2017, 1107-1114. https://doi.org/10.1007/s00170-017-0584-7

[50] Feng L., Wang J., Shi X., Chai C., Applied Physics A, 125, 2019, 261. https://doi.org/10.1007/s00339-019-2562-4

[51] Feng L., Zhao L., Qiang X., Liu Y., Sun Z., Wang B., Applied Physics A, 119, 2015, 75-83. https://doi.org/10.1007/s00339-014-8959-1

[52] Haryono M.B., Lin K.W.Y., Thant K.K.S., Subannajui K., Waritanant T., Journal of Physics - Conference Series, 2696, 2024, 012001. https://doi.org/10.1088/1742-6596/2696/1/012001

[53] Kan T., Xu J., Xie J., Journal of Physics - Conference Series, 2230, 2022, 012027. https://doi.org/10.1088/1742-6596/2230/1/012027

[54] Cao M., Li K., Dong Z., Yu C., Yang S., Song C., Liu K., Jiang L., Advanced Functional Materials, 25[26] 2015, 4114-4119. https://doi.org/10.1002/adfm.201501320

[55] von Rohr P.R., Poulikakos D., Stamatopoulos C., Milionis A., Ackerl N., Donati M.A., de la Vallée P.L., ACS Nano, 14[10] 2020, 12895-12904. https://doi.org/10.1021/acsnano.0c03849

[56] Zhang Y., Cao M., Peng Y., Jin X., Tian D., Liu K., Jiang L., Advanced Functional Materials, 28[5] 2018, 1704220. https://doi.org/10.1002/adfm.201704220

[57] Grounds A., Still R., Takashina K., Scientific Reports, 2, 2012, 720. https://doi.org/10.1038/srep00720

[58] Li K., Zeng X., Li H., Lai X., Applied Surface Science, 346, 2015, 458-463. https://doi.org/10.1016/j.apsusc.2015.03.130

[59] Zhou W., Yang F., Yuan L., Diao Y., Jiang O., Pu Y., Zhang Y., Zhao Y., Wang D., Materials,15, 2022, 8634. https://doi.org/10.3390/ma15238634

[60] Song H.J., Shen X.Q., Ji H.Y., Jing X.J., Applied Physics A, 99, 2010, 685-689. https://doi.org/10.1007/s00339-010-5593-4

[61] Jiang W., Mao M., Qiu W., Zhu Y., Liang B., Industrial and Engineering Chemical Research, 56, 2017, 907-919. https://doi.org/10.1021/acs.iecr.6b03936

[62] Wu Y., Saito N., Nae F.A., Inoue Y., Takai O., Surface Science, 600[18] 2006, 3710-3714. https://doi.org/10.1016/j.susc.2006.01.073

[63] Zuo S., Xu Z., Yi P., Qiu D., Peng L., Advanced Functional Materials, 2025, e10157.

[64] Mohammed M.G., Sundaresan R., Dickey M.D., ACS Applied Materials and Interfaces, 7[41] 2015, 23163-23171. https://doi.org/10.1021/acsami.5b06978

[65] Mirzadeh M., Dehghani K., Rezaei M., Mahidashti Z., Colloids and Surfaces A, 583, 2019, 123971. https://doi.org/10.1016/j.colsurfa.2019.123971

[66] Sun Y., Wang L., Gao Y., Guo D., Applied Surface Science, 324, 2015, 825-830. https://doi.org/10.1016/j.apsusc.2014.11.047

[67] Tsai B.F., Chen Y.C., Ou S.F., Wang K.K., Hsu Y.C., Internatioal Journal of Applied Ceramics Technology, 16[1] 2019, 211-220. https://doi.org/10.1111/ijac.13104

[68] Abbasi S., Nouri M., Rouhaghdam A.S., Thin Solid Films, 762, 2022, 139541. https://doi.org/10.1016/j.tsf.2022.139541

[69] Gao Y., Sun Y., Guo D., Applied Surface Science, 314, 2014, 754-759. https://doi.org/10.1016/j.apsusc.2014.07.059

[70] Li S.Y., Li Y., Wang J., Nan Y.G., Ma B.H., Liu Z.L., Gu J.X., Chemical Engineering Journal, 290, 2016, 82-90. https://doi.org/10.1016/j.cej.2016.01.014

[71] Xiang G.X., Li S.Y., Song H., Nan Y.G., Microelectronic Engineering, 233, 2020, 111430. https://doi.org/10.1016/j.mee.2020.111430

[72] Hu L., Zhang L., Wang D., Lin X., Chen Y., Colloids and Surfaces A, 555, 2018, 515-524. https://doi.org/10.1016/j.colsurfa.2018.07.029

[73] Zhang Y., Chen G., Wang Y., Zou Y., Surface Review and Letters, 28[5] 2021, 2150027. https://doi.org/10.1142/S0218625X2150027X

[74] Zhu M., Tang W., Huang L., Zhang D., Du C., Yu G., Chen M., Chowwanonthapunya T., Materials, 10, 2017, 628. https://doi.org/10.3390/ma10060628

[75] Fadeeva E., Truong V.K., Stiesch M., Chichkov B.N., Crawford R.J., Wang J., Ivanova E.P., Langmuir, 27[6] 2011, 3012-3019. https://doi.org/10.1021/la104607g

[76] Manoj T.P., Rasitha T.P., Vanithakumari S.C., Anandkumar B., George R.P., Philip J., Applied Surface Science, 512, 2020, 145636. https://doi.org/10.1016/j.apsusc.2020.145636

[77] Rasitha T.P., Thinaharan C., Vanithakumari S.C., Philip J., Colloids and Surfaces A, 636, 2022, 128110. https://doi.org/10.1016/j.colsurfa.2021.128110

[78] Rasitha T.P., Philip J., Applied Surface Science, 585, 2022, 152628. https://doi.org/10.1016/j.apsusc.2022.152628

[79] Gao X., Tong W., Ouyang X., Wang X., RSC Advances, 5, 2015, 84666-84672. https://doi.org/10.1039/C5RA15293C

[80] Zhang X., Wan Y., Ren B., Wang H., Yu M., Liu A., Liu Z., Micromachines, 11, 2020, 316. https://doi.org/10.3390/mi11030316

[81] Yang Z., Zhu C., Zheng N., Le D., Zhou J., Materials, 11, 2018, 2210. https://doi.org/10.3390/ma11112210

[82] Wang Y., Chen J., Yang Y., Liu Z., Wang H., He Z., Nanomaterials 2022, 12, 2086. https://doi.org/10.3390/nano12122086

[83] Wang L., Li H., Song J., Sun Y., Surface and Coatings Technology, 302, 2016, 507-514. https://doi.org/10.1016/j.surfcoat.2016.06.057

[84] Long D.P., Xue J.R., Yan Z.X., Advanced Materials Research, 834-836, 2014, 29-32. https://doi.org/10.4028/www.scientific.net/AMR.834-836.29

[85] Shen Y., Tao J., Tao H., Chen S., Pana L., Wang T., Soft Matter, 11, 2015, 3806-3811. https://doi.org/10.1039/C5SM00024F

[86] Zhang S., Zhang Y., Liu M., Wang B., Liu P., Bai X., Cui C., Qu L., New Journal of

Chemistry, 45[38] 2021, 17862-17870. https://doi.org/10.1039/D1NJ02998C

[87] Zhang Y., Gong A., Wang T., Zhang S., Colloids and Surfaces A, 717, 2025, 136776. https://doi.org/10.1016/j.colsurfa.2025.136776

[88] Meng M., Chai B., Yang K., Xiong J., Luo Y., Liang Y., Zhang Z., Liu C., Nano Letters, 25[7] 2025, 3011-3019. https://doi.org/10.1021/acs.nanolett.5c00139

[89] Bala M., Singh V., Journal of Molecular Liquids, 394, 2024, 123780. https://doi.org/10.1016/j.molliq.2023.123780

[90] Hall L.S., Hwang D., Chen B., van Belle B., Johnson Z.T., Hondred J.A., Gomes C.L., Bartlett M.D., Claussen J.C., Nanoscale Horizons, 6[1] 2021, 24-32. https://doi.org/10.1039/D0NH00376J

[91] Qian C., Chen Z., Meng X., Li Q., Chen X., Chemical Engineering Journal, 469, 2023, 143819. https://doi.org/10.1016/j.cej.2023.143819

[92] Chandesris B., Soupremanien U., Dunoyer N., Colloids and Surfaces A, 2013, 126-135. https://doi.org/10.1016/j.colsurfa.2013.05.002

[93] Ozbay R., Kibar A., Choi C., Advances in Contact Angle, Wettability and Adhesion, 2, 2015, 149-164. https://doi.org/10.1002/9781119117018.ch6

[94] Park S., Back S., Kang B., Additive Manufacturing, 94, 2024, 104474. https://doi.org/10.1016/j.addma.2024.104474

[95] Schmitt M., Hempelmann R., Ingebrandt S., Munief W.M., Gross K., Grub J., Heib F., Journal of Adhesion Science and Technology, 29[17] 2015, 1796-1806. https://doi.org/10.1080/01694243.2014.976000

[96] Yao X., Bai H., Ju J., Zhou D., Li J., Zhang H., Yang B., Jiang L., Soft Matter, 8[22] 2012, 5988-5991. https://doi.org/10.1039/c2sm25153a

[97] Chaudhury M.K., Whitesides G.M., Science, 256, 1992, 256, 1539-1541. https://doi.org/10.1126/science.256.5063.1539

[98] Bain C.D., Burnett-Hall G.D., Montgomerie R.R., Nature, 372, 1994, 414-415. https://doi.org/10.1038/372414a0

[99] Schmitt M., Gross K., Grub J., Heib F., Journal of Colloid and Interface Science, 447, 2014, 229-239. https://doi.org/10.1016/j.jcis.2014.10.047

[100] Guo Q., Zhan F., Li M., Shi Y., Wen J., Zhang Q., Zhou N., Wang L., Mao H., Surfaces and Interfaces, 56, 2025, 105623. https://doi.org/10.1016/j.surfin.2024.105623

Materials Research Forum LLC
https://doi.org/10.21741/9781644903896

[101] Wang, Y., Tao, X., Tao, R., Zhou, J., Zhang, Q., Chen, D., Jin, H., Dong, S., Xie, J., Fu, Y.Q., Sensors and Actuators A, 306, 2020, 111967. https://doi.org/10.1016/j.sna.2020.111967

[102] Wang, Y., Tao, R., Zhang, Q., Chen, D., Chen, X., Li, D., Fu, Y.Q., Xie, J., Proceedings of the IEEE International Conference on Micro Electro Mechanical Systems, 2020, 1126-1129. https://doi.org/10.1109/MEMS46641.2020.9056176

[103] Lin M., Yu C., Hu Y.C., Cheng S., Hu H., Annual International Conference of the IEEE Engineering in Medicine and Biology-Proceedings, 2005, 530-533.

[104] Banuprasad T.N., Vinay T.V., Subash C.K., Varghese S., George S.D., Varanakkottu S.N., ACS Applied Materials and Interfaces, 9[33] 2017, 28046-28054. https://doi.org/10.1021/acsami.7b07451

[105] Wang F., Liu M., Liu C., Zhao Q., Wang T., Wang Z., Du X., Science Advances, 8[27] 2022, abp9369. https://doi.org/10.1126/sciadv.abp9369

[106] Wang R., Jin F., Li Y., Yu X., Lai H., Liu Y., Cheng Z., ACS Applied Materials and Interfaces, 14[51] 2022, 57399-57407. https://doi.org/10.1021/acsami.2c17848

[107] Tian Y., Wang H., Tian Y., Zhu X., Chen R., Ding Y., Liao Q., Applied Physics Letters, 123[6] 2023, 064102. https://doi.org/10.1063/5.0159239

[108] Hirai Y., Mayama H., Matsuo Y., Shimomura M., ACS Applied Materials and Interfaces, 9[18] 2017, 15814-15821. https://doi.org/10.1021/acsami.7b00806

[109] He M., Ding Y., Chen J., Song Y., ACS Nano, 10[10] 2016, 9456-9462. https://doi.org/10.1021/acsnano.6b04525

[110] Chu L., Li W., Zhan Y., Amirfazli A., Advanced Engineering Materials, 25[8] 2023, 2201352. https://doi.org/10.1002/adem.202201352

[111] Bovero E., Krahn J., Menon C., Journal of Bionic Engineering, 12[2] 2015, 270-275. https://doi.org/10.1016/S1672-6529(14)60119-0

[112] Li H., Yang Y., Zhu X., Ye D., Yang Y., Wang H., Chen R., Liao Q., Soft Matter, 19[38] 2023, 7323-7333. https://doi.org/10.1039/D3SM00887H

[113] Tian L., Dou H., Shao Y., Yi Y., Fu X., Zhao J., Fan Y., Ming W. Ren L.Q., Chemical Engineering Journal, 456, 2023, 141093. https://doi.org/10.1016/j.cej.2022.141093

[114] Zhang J., Han Y., Langmuir, 23[11] 2007, 6136-6141.

https://doi.org/10.1021/la063376k

[115] Brunet P., Eggers J., Deegan R.D., Physical Review Letters, 99, 2007, 144501. https://doi.org/10.1103/PhysRevLett.99.144501

[116] Zamboni R., Sebastián-Vicente C., Sadasivan A., García-Cabañes Á., Carrascosa M., Imbrock J., Journal of Colloid and Interface Science, 698, 2025, 137976. https://doi.org/10.1016/j.jcis.2025.137976

[117] Gao C., Wang L., Lin Y., Li J., Liu Y., Li X., Feng S., Zheng Y., Advanced Functional Materials, 28[35] 2018, 1803072. https://doi.org/10.1002/adfm.201803072

[118] Wang L., Zhang C., Wei Z., Xin Z., ACS Nano, 18[1] 2024, 526-538. https://doi.org/10.1021/acsnano.3c08368

[119] Duran J., Physical Review Letters, 87[25] 2001, 254301. https://doi.org/10.1103/PhysRevLett.87.254301

[120] Caballero L.C., Melo F., Physical Review Letters, 93[25] 2004, 258001. https://doi.org/10.1103/PhysRevLett.93.258001

[121] Brunet P., Eggers J., Deegan R.D., European Physics Journal - Special Topics, 166, 2009, 11-14. https://doi.org/10.1140/epjst/e2009-00870-6

[122] Benilov E.S., Billingham J., Journal of Fluid Mechanics, 674, 2011, 93-119. https://doi.org/10.1017/S0022112010006452

[123] Noblin X., Kofman R., Celestini F., Physical Review Letters, 102, 2009, 194504. https://doi.org/10.1103/PhysRevLett.102.194504

[124] Daniel S., Chaudhury M.K., De Gennes P.G., Langmuir 21, 2005, 4240-4248. https://doi.org/10.1021/la046886s

[125] Savva N., Kalliadasis S., Journal of Fluid Mechanics, 754, 2014, 515-549. https://doi.org/10.1017/jfm.2014.409

[126] Bradshaw J.T., Billingham J., Physical Review E, 93[1] 2016, 013123. https://doi.org/10.1103/PhysRevE.93.013123

[127] Ding H., Zhu X., Gao P., Lu X., Journal of Fluid Mechanics, 835, 2018, R1. https://doi.org/10.1017/jfm.2017.824

[128] Guo J., Chen X., Shui L., Physics of Fluids, 32[3] 2020, 031701. https://doi.org/10.1063/1.5143874

[129] Zhu X., Wang H., Liao Q., Ding Y., Gu Y., Experimental Thermal and Fluid

Science, 33[6] 2009, 947-954.
https://doi.org/10.1016/j.expthermflusci.2009.02.009

[130] Chung D.C.K., Katariya M., Huynh S.H. Cheong B.H.P., Liew O.W., Muradoglu M.S., Ng T.W., Colloids and Interface Science Communications, 6, 2015, 1-4. https://doi.org/10.1016/j.colcom.2015.06.001

[131] Li Q., Kang Q., François M., Hu A., Soft Matter, 12[1] 2015, 302-312. https://doi.org/10.1039/C5SM01353D

[132] Datta S., Das A.K., Das P., Lecture Notes in Mechanical Engineering, 2017, 1305-1313. https://doi.org/10.1007/978-81-322-2743-4_124

[133] Tan J., Fan Z., Zhou M., Liu T., Sun S., Chen G., Yongchen S.C., Wang Z., Jiang D., Advanced Materials, 36[24] 2024, 2314346. https://doi.org/10.1002/adma.202314346

[134] Lv C., Zhou T., Liu Y., Zhang L., Zhao H., Si B., Langmuir, 41[13] 2025, 8934-8950. https://doi.org/10.1021/acs.langmuir.5c00256

[135] Savva N., Kalliadasis S., Journal of Fluid Mechanics, 725, 2013, 462-491. https://doi.org/10.1017/jfm.2013.201

[136] Datta S., Das A.K., Das P.K., Langmuir, 31[37] 2015, 10190-10197. https://doi.org/10.1021/acs.langmuir.5b02184

[137] Bradshaw J.T., Billingham J., Journal of Fluid Mechanics, 840, 2018, 131-153. https://doi.org/10.1017/jfm.2018.71

[138] Liu H., Zheng S., Yang X., Liao W., Wang C., Miao W., Tang J., Wang D., Tian Y., ACS Applied Materials and Interfaces, 11[50] 2019, 47642-47648. https://doi.org/10.1021/acsami.9b18957

[139] Bonart H., Kahle C., SIAM Journal on Control and Optimization, 59[2], 2021, 1057-1082. https://doi.org/10.1137/20M1317773

[140] Zhan Z., Wang Z., Xie M., Chen Y. Duan, H., Advanced Functional Materials, 34[1] 2024, 2304520. https://doi.org/10.1002/adfm.202304520

[141] Mo J., Huang H., Wang C., Liang J., Li Z., Liu X., Physics of Fluids, 36[11] 2024, 112021. https://doi.org/10.1063/5.0231440

[142] Chen K., Wu Y., Huang Y., Hsu C., Shieh J., Langmuir, 40[25] 2024, 13219-13226. https://doi.org/10.1021/acs.langmuir.4c01366